Building an API Product

Design, implement, release, and maintain API products
that meet user needs

Bruno Pedro

Building an API Product

Group Product Manager: Kunal Sawant

Senior Editor: Kinnari Chohan

Technical Editor: Vidhisha Patidar

Copy Editor: Safis Editing

Project Coordinator: Prajakta Naik

Indexer: Subalakshmi Govindhan

Production Designer: Prashant Ghare

Marketing DevRel Coordinator: Sonia Chauhan

First published: January 2024

Production reference: 2130224

Published by Packt Publishing Ltd.

Grosvenor House

11 St Paul's Square

Birmingham

B3 1RB, UK

ISBN 978-1-83763-044-8

www.packtpub.com

To my wife, Vânia, and my two sons, Bernat and Enric. Without their ongoing love and support, I wouldn't have been able to write this book.

Contributors

About the author

Bruno Pedro is a computer science professional with over 25 years of experience in the industry. Throughout his career, he has worked on a variety of projects, including internet traffic analysis, API backends and integrations, and web applications. He has also managed teams of developers and founded several companies, including Tarpipe, an iPaaS, in 2008, and the API Changelog in 2015. In addition to his work experience, Bruno has also made contributions to the API industry through his written work, including two published books on API-related topics and numerous technical magazine and web articles. He has also been a speaker at numerous API industry conferences and events since 2013.

About the reviewers

David Roldán Martínez is currently the Head of APIs at Shaper by atmira. He also holds the position of Associate Professor at the Universitat Politècnica de València (Spain) and is actively involved as a researcher at VRAIN (Valencian Research of Artificial Intelligence Network).

Professionally, David is a Business and Solutions architect with over 25 years of experience in software systems architecture. He holds a Ph.D. in Telecommunications Engineering and has a strong educational background, including master's degrees and extensive technological training. Additionally, he is a scientific-technical evangelist, having authored more than thirty books.

David's areas of expertise encompass APIs and their applications in various markets such as Banking, Insurance, and Retail. He is well-versed in Artificial Intelligence, Digital Transformation, and the API and Open Economy.

Christos Gkoros is a seasoned software engineer and architect with over 13 years of experience in the industry. He has worked on a variety of projects in different technologies and industries, always striving to find the best possible solution. With a focus on APIs, he is currently exploring ways to help engineers in areas like API Design, API Management, and Strategy.

Christos has a proven track record of transforming complex challenges into streamlined, secure systems, having spearheaded API design at Postman and microservice architecture at Vodafone. He is passionate about mentorship and education and is committed to helping future talent grow and succeed in the field.

Table of Contents

Part 2: Designing an API Product

Part 4: Releasing an API Product

13

14

15

Part 5: Maintaining an API Product

16

Preface

Building an API Product is a comprehensive guide that ranges from the fundamentals of APIs and their inner workings to mastering the steps involved in building successful API products. With this book, you will be able to confidently and effectively create cutting-edge API products that excel in today's competitive market.

Who this book is for

This is a book that helps product managers and software developers navigate the world of APIs to build programmable products. You don't have to be an experienced professional to learn from this book, as long as you have a basic knowledge of internet technologies and an understanding of how users interact with a product.

What this book covers

Chapter 1, What Are APIs?, introduces you to API fundamentals, origins, and types such as REST, gRPC, AMQP, and MQTT.

Chapter 2, API User Experience, explores how the API user experience is vital, second-degree experience, and the impact of friction on success.

Chapter 3, API as a Product, outlines an API as a standalone product, emphasizing business value, monetization options, support, documentation, and crucial security.

Chapter 4, API Life Cycle, provides an overview of the API life cycle stages, covering design, implementation, release, and maintenance, offering an opinionated approach to API product management.

Chapter 5, Elements of API Product Design, introduces you to the key API product design stages, connecting ideation, strategy, definition, validation, and specification, paving the way for an in-depth exploration.

Chapter 6, Identifying an API Strategy, analyses the strategy stage of API design, emphasizing identifying stakeholders, determining business objectives, and understanding user personas and behaviors.

Chapter 7, Defining and Validating an API Design, covers the techniques for defining and validating API design, starting with strategy-derived information and exploring API mocks, UI integration, and stakeholder iteration.

Chapter 8, Specifying an API, guides you on how to select an API architectural type based on behaviors and capabilities, refining the definition with constraints and industry practices and creating a machine-readable representation with governance rules.

Chapter 9, Development Techniques, offers a beginner-friendly guide to API development, covering language and framework selection, code generation from specifications, prototyping, and extending with business logic,

Chapter 10, API Security, explores API security, emphasizing its importance, distinguishing between authentication and authorization, and introducing a security testing technique called fuzzing.

Chapter 11, API Testing, introduces you to API testing methods, covering contract testing to ensure specification compliance, performance testing execution, and the connection of acceptance testing to user personas.

Chapter 12, API Quality Assurance, covers API quality assurance, introducing behavioral testing to validate against identified behaviors and setting up API monitors for periodic testing.

Chapter 13, Deploying the API, provides an overview of the API deployment process, covering continuous integration, agility, automated testing, deployment, and API gateway trade-offs.

Chapter 14, Observing API Behavior, introduces you to API usage analytics, APM, and user feedback analysis to identify and measure important metrics, usage patterns, and behavior.

Chapter 15, Distribution Channels, covers API distribution strategies, including pricing, API portals, marketplace listing, and documentation options to enhance user activation.

Chapter 16, User Support, delves into ways to ensure user success with an API, covering support channels, forums, and prioritizing bug fixes and feature requests from user feedback.

Chapter 17, API Versioning, explores techniques for managing multiple API versions, handling breaking changes effectively, and communicating changes to users using machine-readable methods.

Chapter 18, Planning API Retirement, discusses API retirement, covering its definition, considerations, and communication to users and conducting a retrospective to document what you have learned from the process.

Download the example code files

You can download the example code files for this book from GitHub at `https://github.com/PacktPublishing/Building-an-API-Product`. If there's an update to the code, it will be updated in the GitHub repository.

We also have other code bundles from our rich catalog of books and videos available at `https://github.com/PacktPublishing/`. Check them out!

Conventions used

There are a number of text conventions used throughout this book.

`Code in text`: Indicates code words in text, database table names, folder names, filenames, file extensions, pathnames, dummy URLs, user input, and Twitter handles. Here is an example: "Mount the downloaded `WebStorm-10*.dmg` disk image file as another disk in your system."

A block of code is set as follows:

```
html, body, #map {
  height: 100%;
  margin: 0;
  padding: 0
}
```

When we wish to draw your attention to a particular part of a code block, the relevant lines or items are set in bold:

```
[default]
exten => s,1,Dial(Zap/1|30)
exten => s,2,Voicemail(u100)
exten => s,102,Voicemail(b100)
exten => i,1,Voicemail(s0)
```

Any command-line input or output is written as follows:

```
$ mkdir css
$ cd css
```

Bold: Indicates a new term, an important word, or words that you see onscreen. For instance, words in menus or dialog boxes appear in **bold**. Here is an example: "Select **System info** from the **Administration** panel."

> **Tips or important notes**
> Appear like this.

Get in touch

Feedback from our readers is always welcome.

General feedback: If you have questions about any aspect of this book, email us at `customercare@packtpub.com` and mention the book title in the subject of your message.

Errata: Although we have taken every care to ensure the accuracy of our content, mistakes do happen. If you have found a mistake in this book, we would be grateful if you would report this to us. Please visit www.packtpub.com/support/errata and fill in the form.

Piracy: If you come across any illegal copies of our works in any form on the internet, we would be grateful if you would provide us with the location address or website name. Please contact us at copyright@packtpub.com with a link to the material.

If you are interested in becoming an author: If there is a topic that you have expertise in and you are interested in either writing or contributing to a book, please visit authors.packtpub.com.

Share Your Thoughts

Once you've read *Building an API Product*, we'd love to hear your thoughts! Scan the QR code below to go straight to the Amazon review page for this book and share your feedback.

https://packt.link/r/1837630445

Your review is important to us and the tech community and will help us make sure we're delivering excellent quality content.

Download a free PDF copy of this book

Thanks for purchasing this book!

Do you like to read on the go but are unable to carry your print books everywhere?

Is your eBook purchase not compatible with the device of your choice?

Don't worry, now with every Packt book you get a DRM-free PDF version of that book at no cost.

Read anywhere, any place, on any device. Search, copy, and paste code from your favorite technical books directly into your application.

The perks don't stop there, you can get exclusive access to discounts, newsletters, and great free content in your inbox daily

Follow these simple steps to get the benefits:

1. Scan the QR code or visit the link below

https://packt.link/free-ebook/9781837630448

2. Submit your proof of purchase
3. That's it! We'll send your free PDF and other benefits to your email directly

Part 1:
The API Product

This part provides a comprehensive guide to API development and management, beginning with fundamental concepts, types, and origins, followed by a focus on user experience and the significance of the API as a standalone product. It then delves into the API life cycle stages, covering design, implementation, release, and maintenance, with an opinionated approach for effective API product management.

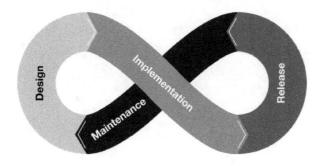

In this part, you'll find the following chapters:

- *Chapter 1, What Are APIs?*
- *Chapter 2, API User Experience*
- *Chapter 3, API-as-a-Product*
- *Chapter 4, API Life Cycle*

1

What Are APIs?

APIs are the most powerful technology available today. While the API acronym can be deceitfully simple, the concept it describes offers infinite possibilities. Welcome to the world of APIs, where you'll learn how to build an API product. Your first step in this expedition is to first learn what an API is. In this chapter, you will understand the nature of APIs, looking back to their origins. You'll also get to know which technologies and tools are available for you to use.

The chapter begins by exploring different types of networks, such as the internet, and how APIs work on them. You will then be guided through the history of APIs. You'll see how APIs came to life and understand how certain concepts in use today were born. Finally, you'll see that there are different technologies and tools that you can use to build an API product from scratch.

By the end of this chapter, you will know that APIs can exist on different types of networks. You will understand what those networks are and what the most appropriate one for your API product is. You will also know that there are synchronous and asynchronous APIs and what those terms mean. Most importantly, you will know how to pick the right type of API and tools to build your API.

In this chapter, we're going to cover the following main topics:

- The different types of APIs
- The history of APIs
- Available technologies and protocols

The different types of APIs

This section gives you an overview of the different types of APIs that exist. APIs are split between local and remote, and then by the protocols that they adhere to. You'll start by understanding what an API is at a high level and why it's so important. Then, you'll dive into the different types of APIs. Let's get started.

Application programming interfaces, or **APIs**, allow applications to be used programmatically. They create an interface—a layer of abstraction—that opens applications to interactions from the outside. The interface has the goal of standardizing any connection to the application. Suppose you think about an interface as a common boundary between two entities. In that case, an API is a way to let an application communicate with other entities in a programmatic fashion.

This type of interaction is what you can call an integration. Integrating different applications, or different parts of the same application, lets you build products by putting together pieces that are ready to be used. Instead of creating all the features of a product from scratch, you can utilize functionality that is already available in the form of an API. That alone is a powerful tool to use. Creating a product by using APIs can be done in a fraction of the time it would take if you develop all the features yourself. That happens because you can reuse pieces of functionality that are standardized and well understood. Those pieces of functionality can be a whole product, a single feature, or a subset of a product represented by a selection of features. Different types of APIs provide different types of interactions, as you're about to learn.

Local APIs

Local interfaces are the most used type of API, even if they're often seen as invisible. All the applications that run on a device need to communicate with the hardware. Applications interact with the device via local APIs. They offer the advantage of providing a standard method of programming the device to behave according to what users want. **POSIX** is one such standard created by the IEEE Computer Society. It stands for **Portable Operating System Interface,** and its goal is to establish a layer of communication that is standard across different operating systems. Another similar standard is the **Single UNIX Specification (SUS). macOS**, a popular operating system developed by Apple Inc., is considered partly compliant with POSIX and fully compliant with the SUS. This means that anyone that interacts with macOS knows that it follows certain rules and conventions that have been standardized. In theory, an application that is built to run on macOS could also run on other systems that are compliant with the same standards.

Another way of introducing a standardized local layer of communication is by using common software libraries. Even if a system doesn't follow a full standard, some of its parts can use standardized libraries. **Java** offers a popular set of libraries and APIs that can be used across systems. The programming language was created by Sun Microsystems—acquired by Oracle Corporation—and has been used on almost all operating systems. Java fully embodies the goal of standardizing how applications interact. Its slogan is *Write once, run anywhere*, and it symbolizes the importance that its creators give to standardized interfaces. Java's versatility is enormous. You can use Java to create mobile applications that run on Android devices, desktop applications, and everything in between.

By now, it's clear that operating systems' standards and libraries offer a way of interacting with the lowest layers of computing devices. Another form of abstraction that encapsulates reusability at a higher level is available through software modules. Most modern scripting programming languages have the ability to create and use modules. Modular software development has become a popular way of building applications. Modules provide functionality that is ready to be used and increase the speed at which applications are built.

A widespread module system exists for the JavaScript programming language. Its name is **npm**, the **Node Package Manager**. Its authors claim that npm is the *world's largest software registry*, with over one million modules available to be used by anyone. According to GitHub, JavaScript is the number one used programming language at the time of writing. In fact, JavaScript has been in the first position for the last eight years at least. Because npm is used by applications written in JavaScript, it's the most used module system.

Other module systems exist for different programming languages, and they all share that they want to facilitate the reuse of functionality and increase the speed of developing software. Python, the second most popular programming language, has **PyPI**, the **Python Package Index**. The third most popular programming language, Java, also has its package system, **Maven**. There are all kinds of modules ready for anyone to use on their applications. The point is that anyone is free to create and publish modules and also to reuse modules that other people have published. Hence, a vast ecosystem of modular software development keeps growing.

While local interfaces deal with the interaction between different parts of a local system, some of those parts let you communicate with the outside world. Communication with remote systems is also abstracted and standardized in the form of APIs. Read on to learn more about the different remote APIs and how they can enhance the features of an application.

Remote APIs

Most people think of APIs as a way to interact with software that is running remotely. They tend to ignore all the local interfaces that you've read about before—not because local APIs aren't important, but because they feel invisible. The opposite happens with remote APIs. Instead of being invisible, remote APIs feel like they're the most critical part of an application. The act of starting a connection to a system that is running on a different part of a computer network feels like something worth paying attention to. Remote connectivity can be split according to the type of network being used. Let's focus on **local area networks** (**LANs**) and **wide area networks** (**WANs**) because that's where most APIs operate.

LANs connect devices that are physically in the same location. The types of applications that exist on a LAN are meant to be accessed exclusively by devices that are connected to the same network. APIs that operate on LANs are typically focused on supporting specific classes of applications and not on providing generic services to consumers. In other words, LAN APIs offer a way for devices to connect to applications running on the same network. As with local APIs, here, the goal is to standardize how the same type of applications communicate on LANs.

Databases are one type of application that is widely used in local networks. The ODBC standard was created to standardize communication between applications and databases. **ODBC** stands for **Open Database Connectivity** and is a standard API for accessing databases. Applications that use ODBC can be ported across different database systems without having to be rewritten. You can, for instance, develop a warehouse stock application that uses the MySQL database system. Suppose that, at some point, you decide to switch to Oracle or some other database system. You don't have to rewrite your application as long as the database system supports ODBC. In the same way, if you decide to change the implementation of your application to a different programming language, you don't have to change the database system. As long as the programming language supports ODBC, you know that you'll be able to interact with your database.

Printing is another popular activity on local networks. As you would expect, there are APIs that standardize the communication between printers and other devices in the same LAN. One such API is the **Line Printer Remote** protocol, or **LPR**. This protocol lets you interact with a printer, programming it to print documents and even changing the configuration options of the target printer. Even though printing happens primarily in LANs, it's a type of application that can also be carried out across the internet. To make communication with printers work easily outside LANs, there is a remote API called **Internet Printing Protocol**, or **IPP**. According to Michael Sweet from Apple Inc., "*at least 98% of all printers sold since 2010 support IPP.*" It became so popular because it offers features such as authentication, access control, and encryption of data transmitted to the printer. And it's not the only API that operates on the internet, as you'll see if you keep reading.

When you hear the term *API*, you immediately think about services that run on the internet. That's because wide access to networked services helped popularize the creation of APIs. Externalizing features of an application feels natural in an environment where all services are connected to the same network. Many times, you can even confuse internet APIs with the services that they expose. We often talk about the API as the offering rather than the interface. That indicates that the internet has contributed to the fragmentation of the types of APIs that are available. There are API types that are best suited for reading data while other types are better used for synchronously storing information, and there are types that work well to asynchronously share information about events.

Let's start by exploring API types that let you easily read data from a remote server on the internet. You can say that the simplest way to read data remotely is to directly access a document. However, you would only call that an API if there were some degree of programmability involved. In other words, when you're directly retrieving a document, you're not sending any parameters to an API. To make it programmable, there has to be something on the server that interprets the request parameters and changes the returned output based on what is being requested. A **remote procedure call**, or **RPC**, is an example of a type of API that lets the requester send parameters to a server and, in return, receive information. In the same way that you can read information with it, you can also use it to store information on a server. In that case, you're sending parameters along with the information that you want to store. Depending on the size of the data—what you call an API payload—you can choose what type of API and which technology to use.

As a rule of thumb, anything that happens on the internet works better with short-lived connections. The internet is an open network. Connections between different points on the internet can change without notice, and that can affect the quality of communication. Communicating asynchronously is also an option. There are types of APIs that focus specifically on letting you share information in an asynchronous way. Usually, these are used for sharing information about events, but also for receiving the result of long-running operations. You make a request that you know is going to take a long time to finish. When the request is completed by the server, it will share the result with you. If availability is what matters the most, then you decide the type of API based on what is reliable most of the time. Many APIs end up running on top of the web because it's the most widely used protocol, and you accept it as having a high resilience. In fact, web APIs are what you'll be working on most of the time when you're building API products.

Web APIs are a type of API that uses the internet and web-specific protocols to communicate. In the same way that the remote APIs that you've read about before make remote resources appear as local, web APIs offer the same functionality for resources available on the web. On the web, it's a common approach to identify the supported media types that can be transferred from the server to the client. That is also the case for web APIs. Two of the most used media types are the **Extensible Markup Language**, or **XML**, and the **JavaScript Object Notation**, or **JSON**. These media types can be easily used and interpreted by API client software. The big advantage of web APIs over other types of remote APIs is the features that the web offers. Just by using the web, you have access to content caching, or the ability to temporarily store responses that can be reused between requests. You also have access to authentication mechanisms that don't require any specific implementation. Finally, you become part of a vast ecosystem of server and client tools that are widely available for anyone to use.

While there are different types of APIs, you're reading this book most probably because you're interested in web APIs. As you've seen before, to most people, APIs are a synonym for something such as a programmable interface running on the web. The API that you'll build will probably run on the web as well, so let's use that as a guide throughout the rest of the book. Keeping in mind that there are several types of APIs, let's focus on how to build web APIs. To get there, let's look now at how APIs came to exist and how they have been evolving.

The history of APIs

By now, you already know that there are different types of APIs that you can use depending on what you're trying to achieve. It's important to know how APIs were created and which events were the key contributors to their evolution. Learning how we got to where we are now is the first step to understanding how to build successful products on top of APIs.

To understand how APIs were invented, let's go back in time to circa 1950. At that time, computing was just getting started, and the first known mention of a software module was made by Herman Goldstine and John von Neumann. The authors were referring to local APIs in the form of modules or, as they called it, "subroutines." In their *Report on the Mathematical and Logical Aspects of an Electronic Computing Instrument* from 1947, they explain what a subroutine is. To the authors, a subroutine is a special kind of routine—a set of program instructions that can be reused. They describe that a subroutine has the purpose of being substituted inside any existing routine. It's clear that the authors were focusing on reusability, as you've read before in this chapter. In fact, from then on, APIs have been used as a way to reduce the amount of programming required to build an application.

However, the work of Goldstine and von Neumann was purely theoretical and couldn't be put into practice at that time. It was only when the **Electronic Delay Storage Automatic Calculator**, or **EDSAC**, was created that software modules were actually used. In 1951, Maurice Wilkes, the creator of EDSAC, was inspired by the work of Goldstine and von Neumann. Wilkes introduced the use of modules as a way to reuse functionality and make it easier to write programs from scratch. Wilkes, jointly with Wheeler and Gill, describes how subroutines could be used in their book *The Preparation of Programs for an Electronic Digital Computer*. In this, they explain that with a library of subroutines, it should be simple to program sophisticated calculations. You'd only have to write a master routine that, in turn, calls the various subroutines where the different calculations are executed.

Still, the first time the term "*application program interface*" (notice that the word used isn't "*programming*") appeared was probably in the *Data Structures and Techniques for Remote Computer Graphics* paper published in 1968. In this, its authors, Ira Walter Cotton and Frank S. Greatorex, notice the importance of the reusability of code in the context of hardware replacement. The paper mentions that even if you change the hardware, a "*consistent application program interface could be maintained.*"

Since then, the concept of reusability through the isolation of code to create APIs has been evolving. While initially, the focus was on APIs that could be used locally within an operating system, at a later stage, remote APIs were explored, most notably web APIs. In 2000, Roy Fielding published an intriguing Ph.D. dissertation titled *Architectural Styles and the Design of Network-based Software Architecture*. In it, Fielding analyzes the differences between local APIs—which are based on libraries or modules—and remote APIs. The author calls local APIs library-based and remote APIs network-based. While local APIs have the goal of offering entry points to code that can be reused, remote APIs aim to enable application interactions. According to Fielding, the only restriction that remote APIs have is the ability to read and write to the network where they operate. Let's continue our exploration through the history of APIs. Keep reading to understand how one of the most popular operating systems is in the origins of web APIs.

Unix

Web APIs originated with the Unix operating system and its way of letting different applications—or processes—communicate with each other. In reality, Unix is not a single operating system. It's a group of operating systems with the same root: AT&T Unix. The first version was created in the 1970s by Ken Thompson and Dennis Ritchie at the Bell Labs research center. AT&T decided to license Unix to other companies after the first version was released to the public in 1973. The licensing of Unix made it—and all its variants—one of the most used types of operating systems of all time. One of those variants, the Sun Microsystems Solaris operating system, has contributed the most to the history of web APIs.

From its inception, Unix has been recognized for its modular structure, where individual processes are created with simplicity in mind and aimed at seamless collaboration. This approach, referred to as the Unix philosophy, has become one of the primary reasons for its immense success and a crucial factor in the evolution of web APIs. Brian Kernighan, one of the developers of the Unix operating system, described interoperability during a demo performed in 1982:

> *"(...) you can take a bunch of programs and stick them together end-to-end so the data simply flows from the one on the left to the one on the right."*

Inter-process communication, or **IPC**, is a system integrated into Unix that enables the transfer of messages between various processes. IPC is a collection of APIs that allows developers to coordinate the execution of concurrent processes. The IPC framework offers multiple forms of communication, including pipes, message queues, semaphores, shared memory, and sockets, to accommodate the needs of diverse applications. However, it's worth noting that, except for sockets, all other communication methods are confined to processes running on the same server, limiting their scope and functionality. It's precisely with sockets that network APIs gained traction and different use cases emerged.

Network APIs

Sun Microsystems leveraged the functionality of network sockets to introduce a method of communicating with remote processes, known as **RPC**. The concept of RPC was first introduced in the 1980s as part of Sun's **Network File System (NFS)** project and adhered to the calling conventions used in Unix and the C programming language. It rapidly gained popularity due to its ability to enable any running application to send a request over a network to another application, which would then respond with the result of the requested operation. The messages and responses are encoded using the **External Data Representation** format, or **XDR**, which provides a standard format understood by both the producer and consumer. The RPC protocol offers the capability to deliver messages with XDR payloads through either the **User Datagram Protocol (UDP)** or the **Transmission Control Protocol (TCP)**, thereby providing compatibility with different network types. While UDP is a protocol more oriented toward performance, TCP offers more reliability in situations where the quality of the network is questionable.

The journey from the first implementations of RPC to its official publication as an **Internet Engineering Task Force (IETF) Request for Comments**, or **RFC**, took about a decade. The RPC protocol was first published as *RFC 1831* in 1995 and underwent various transformations through subsequent versions until it reached its latest form in 2009, as described in *RFC 5531*. That year, Sun Microsystems changed the RPC license to the widely used three-clause BSD, which made it available for free to anyone. Today, most variations of the Unix operating system provide some form of native RPC support. At the same time, Microsoft Windows also offers official support for the protocol through its **Services for UNIX (SFU)** software package. Other operating systems offer RPC compatibility with various implementations for programming languages such as C, C++, Java, and .NET.

Despite RPC's widespread popularity and reputation as a lean and straightforward protocol to implement and use, there may be better choices for heterogeneous network environments. One of the primary issues with RPC has to do with the passing of parameters and the interpretation of data between clients and servers that are written in different programming languages. Although RPC relies on the XDR format to describe payloads, the definition of data types can vary between operating systems and programming languages, resulting in the misinterpretation of information. This has led to the rise of other protocols that provide an abstraction layer for messages and parameters.

The concept of **service-oriented architecture**, or **SOA**, emerged as the prevailing standard for facilitating collaboration among applications operating in heterogeneous environments. Around the same period, various internet-based public services were gaining popularity, with the **World Wide Web (WWW)** particularly attracting the attention of a wider audience outside the academic community.

The web

In 1989, Tim Berners-Lee, an English scientist, created the WWW. Since then, it has become the primary means of accessing information and communicating online. During its early years, the web comprised simple interconnected blocks of information, known as web pages, which could be accessed to view information. These web pages were manually updated by webmasters, who were responsible for maintaining the content. With the rise of commercial web initiatives, various services were developed to allow individuals to upload and share personal information such as photos, blogs, and other forms of multimedia. This led to the creation of desktop applications that enabled users to interact with online services more efficiently. Initially, these applications were used to download information, but eventually, they allowed users to upload content to the web.

The way desktop content-creation applications communicated with the newly launched web services gave rise to what we now refer to as web APIs. One such early example was Flickr, a widely used photo-sharing service that allowed developers to interact with it via a web API. The API enabled developers to perform a range of tasks such as uploading, downloading, listing, and searching photos from a single user account or across the entire service. Business applications such as Salesforce also benefited from what web APIs had to offer. In fact, Salesforce was probably the first modern web API to be launched and used.

On the more closed side of the software industry, other protocols started to emerge, with the goal of simplifying the life of developers and integration designers. One such protocol gained significant popularity because of its natural integration with existing Microsoft tools. It was the **Simple Object Access Protocol**, or, in short, **SOAP**. Microsoft promoted SOAP, and it became the number one way to integrate its different products. SOAP became popular outside of the Microsoft world with support from several programming languages and operating systems. For some time, SOAP was seen as the successor of RPC and the way to connect all kinds of business applications. You can see a simplified illustration of a SOAP request here:

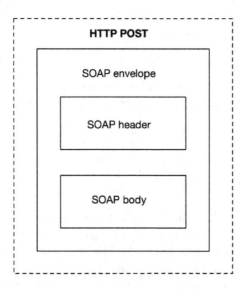

Figure 1.1 – A simplified illustration of a SOAP request

Around the same time, another protocol was being developed to utilize as many features of HTTP as possible and better meet the needs of web services. This resulted in the creation of the **Representational State Transfer Protocol**, commonly known as **REST**. Its creator was Roy Fielding, who you already know from before in this chapter. In his Ph.D. dissertation, Fielding not only described how remote APIs should operate but also invented REST:

REST is a hybrid style derived from several of the network-based architectural styles (...) and combined with additional constraints that define a uniform connector interface.

The network-based architectural styles that Fielding refers to include cache and client-server, two things that are familiar in web interactions. Compared to SOAP, REST is much easier to understand and process and a natural winner on the open web because it doesn't need as many rules to operate as SOAP does. You can see a simple illustration of this protocol here:

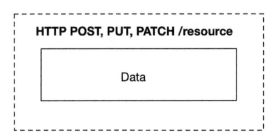

HTTP POST, PUT, PATCH /resource

Data

Figure 1.2 – A simplified illustration of a REST request where data is sent from a client
to a server to create (POST), replace (PUT), or modify (PATCH) a resource

However, because it's more open, it doesn't offer the level of control and security that SOAP offers. While that might not be an issue with non-critical applications, it is a must for business-related APIs. Because of that, another protocol was created. This time, the goal was to increase control over what was transmitted so that the information could be validated in a reproducible way.

Google's Remote Procedure Call, or, in short, **gRPC**, was born in 2015. It started to be used by almost all of Google's open web APIs. gRPC works on top of HTTP/2, the second version of HTTP, offering, among other things, low latency and the ability to stream data between servers and clients. However, the big advantage of gRPC over other protocols is that it uses a strict interface description language. **Protocol Buffers**, or **Protobuf**, is the format used by gRPC to describe the interface and messages shared between clients and servers. Unlike other languages, Protobuf is binary and offers high security and performance. However, Protobuf is oriented toward providing a way to remotely execute code that is available on the API server.

At around the same time, the need for openly querying large amounts of data started to grow. Facebook, a popular social network, pioneered the use of graph databases. It was clear that REST wasn't the best way to access data that was always changing, and gRPC wasn't the answer either. So, GraphQL, a data query and manipulation language, was created. Compared to other API architectures, its main difference is that it allows clients to define the shape of the data that they want to query—that, too, at runtime, in a dynamic way. Even though it doesn't sound like much, this was a big deal because of two factors. On the one hand, it allowed bandwidth-conscious clients—such as mobile phones—to retrieve just the data they needed, saving precious bandwidth. On the other hand, it opened the available data graph to clients, allowing them to openly query all existing data through the API.

Available technologies and protocols

Read on to learn more about the most relevant technologies and protocols that you can use to build API products. It would help if you didn't restrict yourself to using web APIs. Instead, you should build the API product that best reflects the needs of your users, and to do that, the more knowledge you have about what's available, the better. This section offers detailed information about the different API technologies, the communication protocols, and the tools that are considered important to someone building an API product. Let's start by splitting the knowledge into the areas of communication protocols, implementation technologies, and tools.

Communication protocols

Among the available communication protocols, the ones that are most overlooked are the ones that run on local networks. Usually, local network protocols help **Internet of Things** (**IoT**) devices and applications communicate. These protocols use a low amount of power to preserve the batteries of the devices that they support. Among IoT protocols, you have one called **Zigbee**. This protocol is an IEEE *802.15.4*-based specification and is used by popular home automation companies. Philips Hue, for instance, uses Zigbee to power communication between lamps and other devices. Indesit, a house appliance manufacturer, uses Zigbee to adjust its washing machines' cycle starting time according to the price of electricity. Yale, one of the oldest locks companies, uses Zigbee to control its smart door locks. Samsung, a technology manufacturer, uses Zigbee to let you control and monitor its smart refrigerators. If you're thinking about building an API product that interacts with a local device, then a protocol similar to Zigbee might be a good choice.

Even though Zigbee has been gaining popularity, the fragmentation of local connectivity protocols is something to pay attention to. For that reason, a group of organizations created a new standard called **Matter**. Among the organizations behind Matter is, in fact, Zigbee. Matter's goal is to help new product builders adopt a communications standard that they know will work with a vast array of products.

Let's now focus on communication protocols that operate on the internet. While local network protocols solve challenges related to power consumption and interoperability, internet protocols are more focused on the reliability of communication. Reliability, in this case, means the ability to consistently transport information between a user and a server. Users unknowingly engage with servers while they're performing their online activities. The protocol behind most online activities is the **Hypertext Transport Protocol,** or **HTTP**. From a user perspective, it's as if the information is right in front of you, being displayed on the screen of the device that you're using. From a communication perspective, information is traveling across the world using the internet to arrive at your device and then be displayed. HTTP is behind that communication and translates what users request into commands that are sent to servers. The web is powered mostly by HTTP, and when you refer to APIs, most of the time what you're referring to are web APIs.

In summary, HTTP is the protocol behind most of the available APIs. HTTP is a protocol that works in a synchronous way. In other words, when users request something, the information is sent to a server, and the client waits until a response is available. The response might become available almost immediately, in which case the interaction feels like it's happening immediately. Or, in some situations, it might take longer for a server to produce a response. Other protocols have been created to handle situations where the user doesn't need a response immediately or the user doesn't want to wait for a response to become available. Those protocols handle what is called asynchronous communication.

Among the available asynchronous ways of communicating, you have the **Advanced Message Queuing Protocol,** or **AMQP**. This is a protocol that is primarily focused on handling queues of messages. A queue of messages is a group of pieces of information that are stored in a specific order. Each message is picked from the queue and processed one after another. Messages can contain any information and can be used in a variety of patterns that enable multiple kinds of products. You can have messaging patterns that let users perform a command asynchronously. Other patterns are used to let users receive notifications on their devices. There are even patterns that let a server broadcast messages to a group of users without having to connect to each user individually. The important thing to retain about AMQP is that it lets you create interactions that don't require an immediate response from the server.

Another asynchronous protocol that is popular among IoT products is **Message Queuing Telemetry Transport**, or **MQTT**. This protocol focuses on being lightweight in the information it needs to let messages flow from a server to a client. MQTT was built to be as simple as possible and enable low-powered devices to subscribe to information that servers make available. Before, you saw how IoT devices can communicate synchronously inside a local network using Zigbee. In this case, MQTT enables those devices to send and receive information to other devices in an asynchronous way.

Implementation technologies

Now that you know what the available communication protocols are, let's see which technologies you can use to build API products. Starting with local networks, the technologies that you can use either work on top of protocols such as Zigbee or something at a higher level such as HTTP or MQTT.

Let's start with Zigbee. The Zigbee protocol describes three types of devices that can operate on the network. You can build a Zigbee coordinator, a router, or an end device. The Zigbee coordinator manages communication between other types of devices. It can also serve as a bridge between Zigbee and other types of networks, such as the internet. The Zigbee router is responsible for making sure that the information flow reaches all the devices present in the network. Finally, Zigbee end devices are the final nodes in the network. You can build on top of the Zigbee protocol by using—or asking your engineering team to use—one of the available frameworks. The **Connectivity Standards Alliance** (**CSA**), formerly known as the Zigbee Alliance, offers documentation and pointers to implement solutions for Zigbee and also for the Matter protocol. Operating systems such as Tizen offer direct support for Zigbee to applications built to run on it.

Moving into internet technologies, let's look at what you can build on top of HTTP. Anything that runs on top of HTTP is understood as "the web." Fortunately, there are plenty of technologies and approaches to building web APIs. From frameworks to API specifications, you have a lot to choose from. There may be many solutions because the web itself is the most used communication platform. To begin with, most programming languages offer a way of building a web API from scratch. For instance, Node.js, a popular programming language, has the Express.js framework. Python, another language, offers Flask and FastAPI. And there are other options for languages such as Java or PHP. You can pick the language and framework that best fits your needs and where you or the engineers that work with you feel more comfortable.

To specify the API that you're building, you also have different options. In this case, depending on the type of web API that you're building, you have different specification standards. OpenAPI is the preferred specification for REST APIs. While REST restricts your API consumers to the resources and operations that you make available, you can offer more generic access through GraphQL. Essentially, GraphQL lets your API consumers access the data that you are exposing as a graph. In other words, you don't have to provide all the queries and operations beforehand because consumers can navigate the data itself. If you're concerned about performance and data validation, you can use the gRPC specification. With this approach, you have a specific format for the information that is shared between consumers and servers, making the communication stricter than with the REST approach. All these technologies provide a synchronous communication solution. Read on to learn how you can build asynchronous APIs.

If the API that you're building doesn't require an immediate response, or if the operation that you offer takes too long to process, then you can think of offering an asynchronous solution. Asynchronous APIs usually make use of two communication channels. One of the channels is used by the API consumer to send information to the server. The API consumer uses this channel to execute an operation on the server or to request certain information. The second channel is used by the server to communicate the result of the request back to the consumer. Those two channels can coexist on the same type of network or can use totally different approaches. One example of a second channel that runs on top of HTTP is called **Webhooks**.

With Webhooks, you ask the API consumer to provide a way to receive a request from the server. When the server has information to be shared with the consumer, it uses the Webhook URL to push the available data. The consumer then receives the request and the information that they were waiting for. Another way of building an asynchronous API on top of the web is to use something called WebSockets. In this case, the API will be used by web browsers to communicate with the server. The goal is to open a direct communication channel that can be used by both the browser and the server to send information in both directions. This will allow the server to send information to the browser asynchronously. As an example, this is how some solutions, such as browser notifications, are implemented.

If we now move to other asynchronous protocols, there are different products that you can use. RabbitMQ is a product that provides an asynchronous communication broker that runs on the AMQP protocol. Mosquitto is another broker that in turn runs on MQTT. Another product that provides asynchronous messaging solutions is Kafka. Even though Kafka uses its own communication protocol and message format, it's worth mentioning because it's one of the most used asynchronous solutions.

Tools

By now, you know about the protocols, formats, and technologies that are available to help you build an API product. Continue reading to learn about the tools that you can use to get your API up and running. You have tools that let you design and define how your API will behave. Other tools help you validate your API and offer a mockup. There are also tools that convert your API definition into running code that can be deployed to a running server. Whichever tool you use, remember that it doesn't replace your knowledge. You should be able to do things on your own by understanding the principles and the theory behind your actions. Let's start by looking at API design tools. These are usually web applications that help you create an API definition document. These are important to you because they help you build your API product during the first stages of the process.

One of the most popular API design tools is Postman. This is a web and desktop application that can be used individually or by different members of a team to collaborate. Postman offers an interface that lets you create API definitions and automatically runs validation checks to make sure that your API follows industry best practices. With Postman, you can create an API mock from your API definition. You can share a link to the mock with your potential customers and use it for validating your API design. Your API clients can use the mock to try making requests to the API and seeing what it looks like before you actually release any product.

Another tool in the area of API design is Stoplight. The company offers a web and desktop application that lets you design and document your API. Among its features is the possibility of generating and customizing API portals to offer onboarding and documentation to developers. API design is the strong offering of Stoplight. It offers a unique visual approach to designing an API. Instead of typing your OpenAPI definition into a text editor, you can visually configure how your API will behave.

Swagger is probably the oldest API design tool still available. It offers a web application that lets you design your API by writing its OpenAPI definition. Even though it uses a text editor, it automatically renders the result of what you're typing. You can interactively write your API definition and see the results immediately as API documentation. That's an interesting approach because it gives you the perspective of what your API consumers will view.

Another area that is interesting to you is API documentation. API documentation gives you the ability to explain to your users what your API does and how they can use it. There are several tools in this area, ranging from simple documentation generators to more sophisticated API portals. In fact, all the aforementioned API design tools also allow you to publish API documentation. However, there are other tools that are more focused on API documentation.

One such tool is ReadMe. This is a web application that lets you build interactive API documentation. Anyone that accesses your API documentation will be able to interact with the API and engage with you, or your support team, if needed. It also lets you, the API owner, interact with your users by sharing updates whenever something changes. With ReadMe, you don't have to install anything as the tool hosts the whole documentation.

Another tool that lets you build your API portal is Apiable. This lets you create your API documentation and goes further by letting you manage the signup and onboarding process. Apiable also lets you create what it calls "API products." This feature lets you define signup processes and subscription plans for your API. Apiable manages the process so that you can focus on building and maintaining your API.

If you prefer to use an open source tool that you can run on your machine, you can look at Redoc. This is a tool that generates documentation from an OpenAPI definition document. Redoc follows a popular three-panel page layout with features such as a search bar and API request and response examples. You can install it on your machine and run it every time you update your OpenAPI definition.

The tools that I presented here are meant to be examples of what is available. Keep in mind that tools change—what's important is that you stay updated with what exists. No tool can replace your own knowledge and how you're able to build and maintain your API. However, having the right tools at hand makes your job much easier and more pleasant.

Summary

By now, you have a fair understanding of what APIs are, how they evolved, and how their history is connected to the history of computing and the internet. You also know which technologies and tools you can use to build your API product. Let's look back at all the things you learned in this chapter.

You started by understanding the concept of API as a way to connect different pieces of software together, independently of their location or the communication protocol that they're using. Then, you dived into local APIs, which run locally on the device and help the operating system and the applications that run on top of it communicate with each other. After that, you learned the difference between these local APIs and the remote ones that run on networks. Then, you walked through the history of APIs, seeing that, since the beginning, emphasis has been placed on the reusability of software. You saw how different people, such as von Neumann, Fielding, and Berners-Lee, influenced how APIs work and what they do. From there, you went through the existing technologies, protocols, and tools that are available for you to build an API product.

These are some of the concepts that you've learned in this chapter:

- An API is a programmable way of interacting with an application
- APIs offer reusable functionality that reduces the time it takes to build new applications
- Software modules and programming language libraries can be considered APIs
- APIs exist on different types of networks, not just the web
- Different communication protocols provide different features to the APIs that run on them

The following are things to take into account when choosing between a synchronous and an asynchronous API approach:

- The features that different API standards such as **RPC**, **REST**, **gRPC**, and **GraphQL** offer
- Different tools can help get your job done and help you with API design, documentation, validation, testing, and deployment

Thank you for reading this chapter. In the next chapter, you'll focus on **API user experience** or **API UX**. You'll be able to understand how to identify the users of your API and how to make sure they have the best possible experience. Keep reading to learn more about developer experience, API friction, and other topics related to API UX.

2

API User Experience

A product is nothing without its users, and user experience is at the core of building and managing a product that users love. To effectively measure and improve your **API user experience**, or **API UX**, you must start by identifying those users. This chapter explores the different types of users and how they interact with APIs. You'll be able to see the diversity of ways in which APIs can be consumed. You'll also learn what second-degree user experience is and how it can contribute to the success of an API product.

This chapter begins by exploring the top industries where APIs are consumed the most. We will then describe the notion of API personas and the top professional roles in API usage. This chapter dives deep into developers and how their experience of using an API affects their success. You'll then see how the end user experience is tied to APIs and how you can improve it. Finally, we will explore the idea of API friction and how it negatively influences the user experience and the success of the API product.

By reading this chapter, you will know what the different types of API users are and their jobs and challenges. You will understand that developers are the most significant group of API users. You will learn that to have a successful API product, you need to offer a good developer experience. You will also know that API Design can influence the experience of users of applications created on top of an API. By the end of this chapter, you will be able to identify the factors that contribute to API friction and how to mitigate them.

This chapter dives deep into the following topics:

- Who uses APIs?
- Developer experience
- Second-degree user experience
- API friction

Who uses APIs?

APIs aren't used exclusively by developers. More than half of API users have roles unrelated to software development. According to the *2022 State of the API Report* published by Postman, Inc., a survey of about 40,000 API professionals revealed that the most representative non-developer roles sum up more than 20%. Another similar report published by RapidAPI corroborates this by showing that only 80% of the respondents have a developer role. Additionally, API users come from a variety of industries. While technology companies occupy the largest share of API users, other industries, such as education, healthcare, and banking, represent about 20%, according to the reports mentioned previously.

APIs are not something reserved for developers and the technology industry. It's interesting that most people building APIs still focus on developers as their only users. Instead, as an API builder, you should start exploring the other roles that can also be users of your API. By looking at a single type of user, you might miss out on an important share of your potential user base. Keep reading to learn more about how APIs are used across different industries and how each of the roles uses APIs in their day-to-day activities.

Industries

Several professionals in different roles from a vast list of industries use APIs daily. To better understand how they use APIs, let's first focus on a short list of the most representative industries. The industries that you should look closer at, ordered from the one with the least amount of API users to the one with the most, are education, healthcare, banking, and technology. All these industries represent at least 4% of global API usage, with the highest share of more than 50% of global API usage coming from the technology sector. There are undoubtedly other exciting industries. However, those are less representative than the four we're covering here:

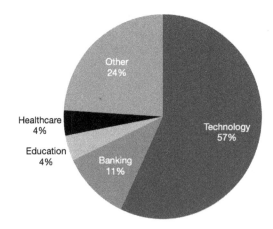

Figure 2.1 – Global API usage breakdown by top industries,
according to 2022 surveys by Postman and RapidAPI

Education

When you mention the word "education," most people picture schools, teachers, and students in uniforms. However, the education industry is much larger than that. Yes, schools are a part of the education industry. Other segments of the education sector include colleges, universities, corporate training companies, and additional specialized training services. Altogether, those segments employ about 5% of the global workforce. That's about 170 million people working in the education industry, representing 4% of the total API usage. If you think about it, it's already a sizeable market. Of course, not all education professionals will become API users. However, it's a ceiling to look at when thinking about your API users.

Education organizations use software – and APIs – in many different ways. The list includes authoring software that helps teachers produce instructional material, reference software such as dictionaries and encyclopedias, proofreading and plagiarism software, and communication tools to help parents and teachers exchange information. As you can imagine, there are already APIs operating in this space. The Oxford Dictionary offers an API that you can use to retrieve definitions of words, synonyms, antonyms, or even sentences where a particular word is being used. Smodin has an API that checks a given piece of content for plagiarism. This type of tool is gaining popularity among education professionals to verify whether student assignments are original. Additio is a company that offers a platform that helps, among others, teachers, students, and parents communicate effectively. All these APIs can be used directly. However, that's not how they're usually accessed in the context of an educational organization.

Most schools and other educational organizations use what are called "learning environments" in the industry. These are the basis of how schools organize themselves and how all the participants interact. Among those environments, the ones that are the most popular are Google Classroom, Blackboard Learn, and Moodle. These **learning management systems**, or **LMSes**, focus on helping users manage aspects of the education program. The features they offer include course management, curricula editing, and attendance monitoring. On top of those specific features, all these systems provide a vast number of integrations in the form of plugins, enhancements, or "applications" that run inside them. And those integrations work because both the learning environments and the third-party software have APIs. By using these learning environments, education professionals and students can interact with APIs indirectly. Another industry where APIs are used mostly indirectly is healthcare. Continue reading to learn more about it.

Healthcare

The healthcare industry is probably the one with the highest chances of surviving. We all, at some point in our lives, will need some type of health assistance. The healthcare sector employs about 2% of the global labor force or around 65 million people, contributing to 4% of the total API usage. Those people are split across different types of healthcare businesses, such as hospitals, health insurance companies, and health equipment. These three segments are attractive because they have to share large amounts of data about patients. Privacy is vital in any business sector. However, healthcare is significantly regulated, so no unwanted patient information is leaked.

The regulation is so tight that there are standards to help companies operate. One of those standards is the **Fast Healthcare Interoperability Resources (FHIR)** standard. It was created by HL7, a non-profit standards development organization. With FHIR, healthcare information can be shared securely and efficiently between operators. FHIR aims to be modular so that information can be produced and consumed granularly. It supports different data formats, such as XML and JSON, and follows existing web standards. The FHIR standard identifies resources that are common in the healthcare industry. Other standards exist to extend and provide more detail to the common resources that FHIR defines. Among those standards is the **United States Core Data for Interoperability (USCDI)**. This standard defines a set of attributes that should be exposed using FHIR at the request of healthcare patients. Those attributes include things such as clinical notes about the patient, information about medication, vital signs, and identification of implantable devices, such as pacemakers. Failure to comply with the existing regulations by not following the correct standards might incur sanctions. Therefore, before building an API in the healthcare sector, it's crucial to have a clear understanding of the regulations and standards.

Let's see how healthcare companies use APIs to share information. One type of information-sharing happens between hospitals and health insurance companies. The most simple form of API exists to help hospitals share medical bills with health insurance companies. Other, more sophisticated use cases include obtaining a patient's medical record to calculate insurance costs, retrieving family medical history to understand the probability of diseases occurring, or calculating health scores based on a patient's visits to healthcare providers. Eligible is a company that offers an insurance billing API, which is used by healthcare applications. OneRecord provides an API that lets operators access a patient's medical records, and 1upHealth provides a health history API. Other types of sophisticated capabilities are also being exposed through APIs. Infermedica, for instance, offers an API that generates a diagnostic based on a patient's symptoms.

Wallgreens has an API that lets healthcare companies order refills of medical prescriptions. The power of all these APIs is that they can be integrated into the management software that hospitals and other healthcare providers use. By integrating with healthcare APIs, the management software can be extended with added features. Banking is another area where the ability to externalize features is essential. Keep reading to see how banks use APIs.

Banking

According to the World Bank's *Global Findex Database*, in 2021, 76% of the world's population had at least one bank account. Banking is a sector that has grown steadily and touched all sectors of society. Estimates indicate that the banking industry employs about 1% of the global workforce or around 35 million people. While 1% seems like a low value, it powers an industry with an estimated market capitalization of over seven trillion US dollars. It also contributes 11% to the API usage of all sectors – all this without mentioning that digital banking has grown steadily. And digital banking is what people use when they access their bank accounts through the web or a mobile application.

While web and mobile access to bank accounts constitute a large share of banking API use, the story doesn't end there. Far from it. The greatest use of banking APIs happens between banks and other financial institutions. Because dealing with money and other financial assets is seen as a critical activity, bank communications are highly regulated. And, as you've noticed with the healthcare sector, whenever there are regulations, standards emerge to fill the void. In the case of banking, there are two primary standards in use: the Open Banking Standard, which was born out of necessity after the European Union revised the **Payment Services Directive (PSD2)** regulation, and the **Banking Industry Architecture Network (BIAN)**. The Open Banking Standard was born in the European Union as a way to unify the PSD2 regulation. It aims to provide guidelines to banking implementors through a set of API specifications and security profiles. The API specifications include payment initiation, both domestic and international, confirmation of funds, recurring payments, and event notifications. It also provides resources for third parties that wish to build mobile and web applications targeting banking customers. The BIAN went beyond Open Banking and offered an ecosystem of different players from the banking sector. Together, they have been offered guides, standards, and enhanced API specifications that conform to international regulations.

Without Open Banking, it wouldn't be possible for people like yourself to build products that interact with banking information. One way that APIs are used in banking is to allow applications to access – and manipulate – consumer information. In 2023, the number of applications that enhance what banks offer has been growing. Goin, for example, is a product that helps people save money and spend better by connecting to their existing bank accounts. Belvo provides a layer of abstraction to let you quickly initiate payments to multiple banks. Brex is a banking service oriented at companies that offers integration with several invoicing and management applications. All these use cases wouldn't be possible without API specifications like the ones promoted by Open Banking and the BIAN. The number of scenarios where banks interoperate with each other has been growing. And at the same time, the technology to support those scenarios has also evolved. Read on to understand how the technology sector not only controls most of the existing APIs but is also their most significant consumer.

Technology

Whenever you think about APIs, you immediately picture technology professionals creating integrations and connecting applications. The technology sector has the most significant share of API consumers. With an estimated 60 million people working in the industry worldwide, or about 2% of the global labor force, technology is the driver of most other business sectors. The technology industry involves companies and professionals working on areas that range from physical communication infrastructure to web start-ups. With the fast pace of digitalization of businesses worldwide, it's easy to understand that software is behind almost everything we do. A famous venture capitalist, Marc Andreessen, once said that "*software is eating the world.*" What he meant is that software removes all the barriers that once existed in everyday transactions. Software can do what used to be done by people in a faster, easier, and cheaper way. And that is primarily because of the existence of APIs.

The technology sector mainly uses APIs to build software that other business industries use. However, it also uses APIs to address its own needs. Technology companies use different types of APIs to manage, control, and maintain the creation and deployment of software. Software development, for instance, usually involves the release of different versions with improvements or fixes over time. The management of those releases is generally done with version control systems. A popular version control system, Git, lets software developers store their work in a repository and distribute the code to other team members. Deploying applications to servers that run on the cloud is also done with the help of APIs. **Amazon Web Services** (**AWS**) offers a vast set of APIs that lets developers provision servers and deploy their applications. New Relic is an observability product that lets developers analyze the performance of their applications by sending telemetry data using an API. All these services have, as their primary customer, the technology industry. In other cases, APIs are built so that different business sectors can use them.

One of the business sectors that consumes a big part of the work of technology professionals is the writing industry. In particular, web writing professionals take advantage of all the APIs that exist to connect various services. Think about blogging software and all the integrations that they provide, or all the tools that surround social media services. WordPress, a popular blogging platform, lists more than 10,000 plugins in its "API" section alone. There are plugins related to marketing and selling products online. Other plugins interact with event management services to let writers display relevant information on their blogs, while others allow writers to interact with emailing services to send messages to their readers periodically. Buffer, one of the most used social media sharing services, can interact with nine different channels. You, as a writer, can share your content across a multitude of social media services all because of APIs. And writing is just one example of a business sector that depends on technology to grow. Writers are one type of professionals that often use APIs without even realizing it. However, other professionals engage with APIs directly. Read on to learn what those groups of professionals, or personas, are and discover their attributes.

Personas

Now that you've seen that APIs are used in several business industries, let's focus on the types of API consumers. According to Postman's *2022 State of the API Report*, other than developers, the more relevant API consumers are business analysts, product managers, students and teachers, software architects, and quality engineers. Together, these groups of professionals constitute 21% of API consumers. As a comparison, developers – full-stack, backend, and frontend – constitute 49% of API consumers.

You can call these consumers personas because each represents a group of people with similar characteristics. According to the Nielsen Norman Group, there are three types of personas: lightweight, qualitative, and statistical. In this section, you'll focus on the lightweight – or proto- – personas. They're called proto-personas because they represent a prototype that you can expand from during your process. The proto-personas that you'll read about have been crafted from my own experience. The way to describe each persona is to identify the attributes that are relevant to you when building an API. You usually care about each persona's jobs, challenges, and tools.

During your API Design process, you're encouraged to perform qualitative persona analysis by conducting user interviews. For now, let's analyze the personas that are the most expected to interact with APIs.

Business analysts

A business analyst is someone that helps organizations manage and improve their internal processes. Analysts obtain and validate business requirements and map them to existing technology. They identify and improve processes while following business orientation. With this information, you can now identify the generic tasks related to API consumption. Business analysts analyze and interpret the usage of tools from existing data, automate the interaction between different tools to improve processes, and obtain summarized information from usage data to present to business leaders. The challenges that a business analyst faces are proving hypotheses to business stakeholders based on existing data and presenting findings from datasets using standard business tools. Business analysts use tools such as spreadsheets, slide presentations, document editors, and data access software during their work. They also engage with the tools that they're analyzing. They consume APIs both directly and indirectly. Direct API consumption happens when they need to access data and system information to help them identify patterns. Indirect API access occurs whenever they set up integrations between two or more existing tools. Finally, through their findings and recommendations, business analysts can influence an organization to build specific APIs.

Product managers

While a business analyst works inside an organization and improves its way of working, a product manager works with potential customers that are not part of the organization. Most of the work of a product manager is around interacting with prospects to identify their needs and translating that evidence into product specifications. Product managers are also owners of the product roadmap, a timeline of all the features proposed to be included in the product. They also spend time presenting the evidence and the roadmap to internal business stakeholders, promoting their ideas of what should be included in the product. Finally, product managers are communicators and often participate in writing engagements where they promote the product and any new features. From an API perspective, product managers analyze existing product usage to identify patterns and present summarized findings to business stakeholders. They engage with existing and potential customers using communication and social media services and publish product promotional information using blogs and social media services. Product managers face the challenges of proving their product feature hypotheses based on their findings, communicating efficiently with users using different channels, and convincing business leaders that their product proposals are worthy of investment. They use tools such as analytics software, spreadsheets, data manipulation software, presentation applications, and blogging and social media services. Most of the tasks involve some form of indirect API consumption. However, product managers also use APIs directly when interacting with data systems. More importantly, product managers are also behind the creation of the APIs that are used by all the other personas.

Students and teachers

Both students and teachers are a part of the education industry, which you learned about in this chapter. Students and teachers are two different groups of people with specific attributes. However, for this examination, we will put them in the same group. To do that, let's focus on the tasks that both students and teachers engage in. They communicate with each other using learning environments and the integrations that are available with other systems. They search for and obtain information about the topics that they study and teach, and they use software to monitor their daily activities. Their challenges are finding the information they need from existing web services and presenting it in a meaningful way. The tools they use include web browsers, document-editing applications, spreadsheets, digital whiteboards, and slide presentation applications. From an API-centric perspective, they're also responsible for creating new concepts and ideas and experimenting with new technology. Take, for example, Roy Fielding, the creator of REST, who you learned about in *Chapter 1*. Fielding shared his ideas about REST in his Ph.D. dissertation. He was a student learning about new API concepts and experimenting. From that learning and experimentation, a new API architecture was born. Speaking of architecture, keep reading to see how software architects interact with APIs.

Software architects

The software architect's role can be seen as a hybrid between the disciplines of design and engineering. Software architects work on understanding what the business goals are, translating that information into actionable technical principles, and designing software that meets the criteria. Software architects can also develop the software themselves or lead a team of developers to do that. Their tasks are then somehow similar to the ones of business analysts but with a focus on software development. They analyze and interpret how people interact with existing software within an organization and design any improvements they feel are needed. To do that, they document the creation of new software and present their proposals to business stakeholders. Software architects' challenges are convincing business leaders of new software design approaches, identifying patterns in software usage to back their hypotheses, and communicating software development plans to team members. The tools that software architects use include code editors, diagram applications, document editors, and slide presentation applications.

Quality engineers

The role of a quality engineer is often related to testing how an application or technology behaves. The discipline of quality assurance has the ultimate goal of increasing customer satisfaction. To achieve that goal, quality engineers must put themselves in the shoes of regular users and implement processes that test the systems in an automated fashion. Quality engineers interact with APIs in situations where the technology they're testing is an API or whenever a test automation benefits from the use of a particular API. In the first case, they use an API directly to test all its capabilities while paying attention to things such as reliability and performance. In the second scenario, they use APIs as a way to automate tests. Products such as Ghost Inspector let quality engineers automate the execution of tests as part of a software build process. If any of the tests fail, the software build is considered a failure, and the new version isn't deployed. Selenium, a web browser automation tool, is among the tools that quality engineers use. Other tools include code editors, version control systems, and presentation applications.

Developers

Developers are at the top of the roles that use APIs directly. This group of professionals uses APIs to decrease the effort it takes to develop software. Instead of developing software features themselves, they can rely on the capabilities of existing APIs to add those features to the applications that they create. Developers do that by integrating their code with other systems and services by connecting to APIs. They also use APIs as a doorway to connect to data sources to avoid accessing databases directly. This approach makes it easier to switch vendors and abstracts the way data is consumed and manipulated. Another use of APIs is to automate recurring tasks. Toil, a term often used in software development, is the accumulation of manual tasks that are not a part of the software being developed. One of the goals of software developers is to eliminate toil by automating tasks as much as possible. Developers use tools such as code editors, version control systems, and API client applications to help them build software.

Developer experience

Now that you've read about the different types of API users and learned that developers are at the top of the list, let's see how their experience can be measured and improved. DX refers to developers' overall satisfaction and productivity when using a technology, product, or platform. This measure of satisfaction is relative to the different tasks that developers perform. In our case, we're interested in the experience that developers have when interacting with APIs. Let's focus on how easy it is to use an API, what documentation and tooling are available, and what the support feels like.

A good developer experience can significantly impact the success of an API. It can increase developer adoption and engagement, improve the quality of the code produced, and reduce the time and effort required to develop applications. On the other hand, a bad developer experience can lead to frustration, decreased productivity, and low developer satisfaction, which can negatively impact the success of the API product. Several factors, or needs, can contribute to how good the developer experience can be. The different needs that developers have are interconnected and have different levels of importance. I call this the "API Hierarchy of Needs." It's a pyramid inspired by the work of the psychologist Abraham Maslow that I created in 2013 (see *Figure 2.2*). While Maslow focused on generic human needs, such as breathing and eating, I'm focusing on the needs of developers while interacting with an API. At the bottom of the pyramid, there's usability, or ease of use, the most critical developer need:

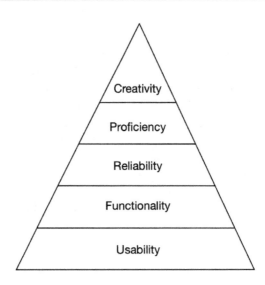

Figure 2.2 – The API hierarchy of needs

The easier an API is to use, the better the developer experience will be. Being easy to use means that you can easily understand what the API does and use it with the tools you're familiar with. The API should be easy to consume and also simple to troubleshoot and test with commonly used debugging and testing tools. When that's not the case, developers need to access documentation to learn how to interact with the API. Offering detailed and accurate documentation is critical to a good developer experience, especially in situations when things don't work as expected. And, in those situations, having access to support channels makes the real difference in terms of experience. There are different types of technical support channels. The vital thing to remember is that support should happen where developers spend most of their time. The easier it is for developers to find the answers to their questions, the better. Support can be public or private. Publicly, you can engage with developers and share information that isn't private. Otherwise, it's better to use private support channels where any sensitive information can be shared between your support team and a developer.

As you can see, without usability, developers won't be able to use your API or even understand how they can get help if they need it. The second layer on the hierarchy of needs is functionality. This means that the API should work as advertised. Working well means that your API should respond to requests as expected and in a performant way. The performance here is not exclusively about speed. It refers to how quickly and efficiently an API can handle client requests and return responses. It is a measure of the API's ability to satisfy the demands placed upon it by the number of consumers and the complexity of their requests. Performance can be affected by various factors, such as the processing power of the server, the efficiency of the API's code, network latency, the volume of data being transferred, and the complexity of the underlying database queries. Performance is often connected to how reliable an API is because it's more complex to maintain both attributes simultaneously.

Reliability, the next layer in the API hierarchy of needs, refers to how the API can consistently deliver correct and expected responses. You can often use reliability as a way to refer to stability. Reliability can then be considered a measure of an API's stability and dependability. To be reliable, your API should be able to handle the expected usage load and still perform as it was designed. There is a trade-off between performance and reliability that has to be considered, depending on the goals you have for the API that you're building. If what matters to you is that requests are handled quickly, and you don't care whether the API fails under a heavy load, then what you consider essential is performance. On the other hand, if you care about knowing that all requests are handled correctly, even if sometimes slowly, then you see reliability as more critical. Ideally, you will find the right balance between performance and reliability.

The fourth layer of the API hierarchy of needs is proficiency. It refers to how well the API is designed and, more importantly, whether it can increase the skills of the developers that use it. In other words, the more proficient an API is, the more the developers that use it can learn and increase their knowledge to do new things. You can see your API as a tool that developers can use to augment their ability to perform their tasks. Proficient APIs are not restrictive in the way developers can use them. Another way to look at proficiency is to measure how flexible an API is. A good developer experience requires an API to be flexible, allowing developers to customize and extend it to meet their specific needs. And by extending the API, developers are increasing their skills.

Finally, let's look at creativity, the top layer of the pyramid. Offering an API that allows creativity means developers can use it in unexpected and creative ways. Creativity opens a range of possibilities never anticipated and can even lead to discovering new features and products. It would be best if you fostered innovation among your API developer community so that they could build new products by mixing different APIs. If you're too rigid about what developers can or cannot do, your API's popularity will certainly decrease. Popularity is usually associated with success, so you'd better pay attention to the API hierarchy of needs if you want your API to succeed.

You now understand the importance of ensuring that your API offers good usability, functionality, and reliability. You also know how vital it is that developers feel proficient and creative when using your API. Even with all these elements in place, many API products need help finding the right audience to be successful. Developer experience is a make-or-break situation for many APIs. However, let's remember that behind every developer, there's an application often used by non-technical users. Now, let's focus on those users and their experience when they indirectly interact with your API.

Second-degree user experience

Second-degree user experience is probably the most overlooked aspect of API experience. While it's easy to understand developer experience, second-degree user experience is something most people don't consider when building an API product. An API with a positive developer experience makes it easy for developers to build applications. In turn, those applications are used by end users who might be someone other than developers. Second-degree user experience is the experience that users have when they interact with the applications that developers build using your API. The interaction with the API is not done directly. Instead, users interact with the API through the application that they're using. That's why we call it second-degree use experience.

The way you design your API can considerably affect the second-degree user experience. Elements such as the choice of API architecture, the authentication scheme, and the way the API returns information to the consumer can directly impact the experience of the users of applications that interact with your API. For instance, whenever you use Twitter, a popular social media application, you may notice that there's no way for you to paginate through the posts. Instead, you see what is called the "infinite scroll," where you have the most recent posts at the top. When you scroll down, the application automatically loads more posts, so you never see an empty list. In other words, you can't jump to a specific page on the list of available Twitter posts. This constraint comes from the Twitter API using token-based pagination, a technique that lets you navigate to the previous and next page of results but never to a specific page. Another example is the commonly used streaming service Netflix. You connect your TV to your Netflix account by typing a code that appears on your TV into your Netflix web account. You sign in to Netflix on the web and then type the code that is shown on your TV. This technique is often called out-of-band authentication and works because Netflix's API provides a mechanism to validate the TV codes. If not, you would have to type your Netflix username and password on your TV, making the process more prone to errors.

As you've seen, second-degree user experience is critical to the success of the applications and services that developers build using your API. One way to make sure that you provide the best second-degree user experience is to talk to developers – your API direct users – to understand what applications they're building. Then, get as much information from second-degree users as possible and design your API while considering their use cases. Not caring about the developer and the second-degree user experience can lead to API friction, as you'll learn in the next section.

API friction

API friction is a concept I derived from David Pogue's article on how technology products have increasingly become more challenging to use. In the *Make Technology—and the World—Frictionless* Scientific American article from 2012, Pogue states that every website where you need to fill in a form or spend time proving that you're a human adds friction. My notion of API friction is a translation of what Pogue describes as product friction seen through the lenses of API users. Even though the result of both is dissatisfaction and a poor experience, the causes of each are not the same.

Previously, you learned about the factors that create a positive developer experience. A straightforward way of understanding API friction is to look at the negatives within developer experience and expand from there. All the challenges, obstacles, and difficulties that developers encounter are a part of API friction. The lower those obstacles are in the API hierarchy of needs, the higher the friction becomes. An obstacle related to API usability creates more friction than one that affects reliability. Let's analyze each of the elements that contribute to increased API friction.

The first obstacle you can think of is the difficulty of understanding what an API is and what it does. If developers need help understanding what an API does, there's no way they'll want to use it. Challenges such as getting to know what you can do with the API or how to start using it are significant sources of frustration. And testing the API is one of the first things that developers do. As a developer, you want to quickly see how you can start working with the API. If you can't do that, you simply walk away. One way to mitigate this obstacle is to prioritize offering your API the best possible onboarding experience. Good onboarding begins with good documentation.

Inadequate or non-existent documentation is obstacle number two that developers face when interacting with your API. Even if you can see what the API does and understand how to work with it, you'll soon need documentation once you dig deeper. And by documentation, I mean a complete reference of everything the API can do and also guides showing how to perform specific use cases. Offering guides is an excellent way to ensure developers can integrate with your API easily and quickly. However, documentation is sometimes not enough to make developers happy. Making sure they can use the API with little effort is undoubtedly a step in that direction.

A **Software Development Kit (SDK)** is a programming library that interacts with an API. The goal of an SDK is to decrease the effort it takes to integrate with an API. Having an SDK that developers can use with their preferred programming language makes your API as easy as calling a local library. Not offering an SDK certainly contributes to an increase in API friction. Even though developers have all the information to use your API, they'll have to write software from scratch to interact with it. That can often lead to spending more time debugging the interaction with your API and is another source of frustration. If debugging doesn't help troubleshoot the problem, developers will need to contact support to find a fix.

Not having a proper API support channel is another obstacle to using your API. Good support means that there's a channel dedicated to API-related questions. The people behind the support channel understand the API, the technology behind it, and the needs of developers. Support requests shouldn't take a long time to be acknowledged and solved. Other complementary forms of support can include discussion forums and chat channels, differentiated support for different SDKs, and even periodic proactive learning sessions.

While these are the most critical factors that contribute to API friction, there are other details that you should pay attention to. API reliability is one of those factors because it's related to how the API behaves repeatedly and consistently. Lack of performance can also generate friction if it's part of the use case of API users. Altogether, anything that reduces the ability to use your API to fulfill your users' objectives contributes to an increase in friction. The way to fix the problem of API friction is to understand your users and their jobs to be done. You can do that by looking at your API from a pure product perspective and treating it as you would treat any other product.

API friction also affects second-degree user experience. The higher the friction that developers feel when interacting with your API, the fewer resources they can spend on understanding how users interact with the applications they're building. API friction makes developers find the path of less resistance, which is often different from the one that results in the best second-degree user experience.

Summary

At this point, you know how to identify the different groups of API users and what matters to them the most. You also understand the different API user experience types and how API friction can translate into an unsuccessful API product. Let's look in detail at the various concepts that you learned by reading this chapter.

You began by learning how to answer this chapter's question: who uses APIs? You dove deep into the different business industries where APIs are consumed, and you understood that the top sectors are education, healthcare, banking, and, of course, technology. Then, you analyzed the different personas that interact with APIs when performing their jobs. You learned that API users are not just developers. From there, you learned about the API hierarchy of needs and how it's used to measure developer experience. On top of that, you understood what second-degree user experience means and what you can do to take it into account when designing your API. Finally, you explored the concept of API friction and how it can negatively influence the success of your API product.

Some of the things you learned in this chapter include the following:

- The groups of individuals who use APIs regularly
- API user personas and their common attributes
- The business industries where APIs are used the most
- The personas that use APIs directly and indirectly
- DX and the factors that influence it
- The API hierarchy of needs and how it's related to developer experience
- The concept of second-degree user experience
- How API design can influence the experience of users of applications built on top of an API
- The factors that generate API friction and how to mitigate them

I hope you enjoyed reading this chapter and what you've learned so far. Now that you know that a good API user experience is crucial to building a successful API product, you're ready to learn more.

The next chapter is about identifying the elements that are part of an API product. Continue reading to learn how to identify the business value of your API and how to monetize it, support it, and ensure that your API-as-a-Product is secure.

3
API-as-a-Product

API-as-a-Product is a term commonly used to identify APIs as products. Instead of viewing an API as a technical solution to integrate systems, you recognize that APIs are themselves products. Products exist to serve users but also to generate value for companies. In this chapter, you'll learn how to identify business value for your API product by employing different monetization models to generate revenue. However, sometimes users need support, and you'll learn how that can become a cost you can't escape.

You'll begin the chapter by learning about the inherent business value that an API product can bring to a company. The chapter continues by analyzing the different options for monetizing an API product and generating revenue. You'll then see why support and documentation are fundamental to the success of an API product. By the end of the chapter, you'll have learned about API security aspects such as logging, monitoring, rate-limiting, and authentication.

After reading this chapter, you will know how to identify and present the business value of an API. You will have learned about the different monetization models and how to implement them. You will understand the role of user support and the various attributes of good API documentation. Finally, you will know the impact that security has on the success of an API product.

This chapter addresses the following topics:

- Business value
- Monetization models
- Support and documentation
- Security

Business value

A business without value has no meaning. Companies exist to provide value to their stakeholders, including shareholders, employees, and also customers. Even companies that claim they're not looking for profit need to generate value to stay alive. In 2022, Yvon Chouinard, the founder of Patagonia, Inc., a popular outdoor clothing company, decided to dedicate all the company's profits to fighting climate change and protecting natural resources worldwide. At the time Chouinard took action, Patagonia's valuation was approximately $3 billion, and the company generated profits of around $100 million a year. The founder started a letter he wrote to Patagonia's employees by saying that the *"earth is our only shareholder"*. Most businesses don't have the chance of doing what Patagonia did, as they can't afford to stop sharing their profits with existing shareholders, investing in new product development, or improving their employees' salaries. The goal, in any case, is to generate long-term value that can be translated into tangible results for a company's stakeholders. According to Marc Goedhart and Tim Koller of McKinsey & Company, *"value-creating companies create more jobs"*. The authors found that there's a correlation between companies with the highest shareholder value and the highest employment growth. Ultimately, a way to define business value is to see it as a perception of the benefit that stakeholders obtain from a company.

Stakeholders can perceive the value that a company creates in different ways. If you're a shareholder, what you care about is the value of the company stock and the dividend amount at the end of the fiscal quarter. If you're an employee, you think about your salary and any bonuses associated with a growth in profit. If you're a customer, you value the quality of the products, their prices, and how the company treats you. Businesses should then look at their stakeholders to understand the kind of business value they should create. Depending on who you want to make happy, you can look at different ways to provide value. Internally, one way to create business value is to look at the company's efficiency and improve it. Efficiency is often associated with the ability to turn a profit. The more efficient a company is, the better it can sell at a lower cost. However, efficiency must be balanced with what employees expect from the company. Otherwise, you end up in a situation where you don't have motivated people to help the company succeed. Speaking of people and happiness, another way to create value is to enhance your company's customer experience. By increasing the satisfaction of your customers, you're making your products more attractive, and more attractive products captivate new customers, generating more sales and more revenue.

For the same reason, business value can also be created by developing new products. Companies that make new products find new selling opportunities and new cohorts of customers. In other words, new products open companies to new markets where new value can be uncovered. Business value is a measure of the success of a company. However, not all business value translates into success in the same way. Therefore, companies periodically measure how different types of value perform so that business direction can be adjusted if needed. Sometimes, companies run experiments to understand how specific actions can lead to more or less business value. Other times, companies need to dramatically change direction by doing what is called "pivoting". The goal is always to find ways to provide gains to the company stakeholders, and one such way is to create a new API product.

The business value that an API generates can be defined by starting with identifying the problem that the API solves. As with any product, an API needs users, and users look for solutions to their problems. Therefore, it's essential to understand who your users are, what their challenges are, and which ones your API can solve. You already know from *Chapter 2* how to identify and categorize different kinds of API users. It's handy to map your potential API users to different personas. Now, you need to understand who will use your API. To do that, you first contact potential users and learn how they would use an API such as yours. Then, you calculate how many users you believe would use your API. With that information, you can estimate the total size of your API user base. However, you still have to understand the value the new API will bring to your company.

One way to understand the value of an API is to investigate the benefits the API will provide to the company. Coming from the identified problems that the API users currently have, you can anticipate and calculate how much users would pay. Having paying users means that your company will obtain an increase in revenue directly generated from the API. Even if users aren't paying for the API, they might be getting more satisfaction just by being able to use it. Increased customer satisfaction is another way to generate business value, as you've seen before in this chapter. Finally, the API might produce an increase in operational efficiency for your company because it can help users and employees automate everyday tasks that once took too much time. In this case, the business value is related to reducing costs, which, given the same revenue, will help the company increase its profits. In any case, the API is considered a cost because it has to be operated. Operational costs for an API product can be split into costs related to payroll, infrastructure, software licenses, marketing, and support. All these costs contribute to the total cost of running an API product. Understanding how much the API costs the company and how that can be mitigated is essential. One figure to consider is the number of API calls needed to achieve break-even—in other words, the number of requests users have to make to your API to get to a point where it becomes self-sustaining. The break-even point is associated with how you plan to monetize your API and how you calculate how much revenue is obtained from its usage. Next, you'll learn how to generate income by applying proven monetization models.

Monetization models

There are many ways in which you can make money with your API. In 2013, John Musser, an API expert, presented 20 possible business models that APIs can follow. According to Musser, API business models evolved from four basic categories. APIs can be free to use, developers can pay for usage, companies can pay developers, or businesses can generate revenue indirectly from usage. The critical takeaway from Musser's presentation is that there is more than one way to generate value from an API and break even. I went through all those 20 business models and summarized them into the three most important categories to understand in the context of API-as-a-Product. The main characteristic of an API-as-a-Product is that the API is the only product being offered. In other words, the API is not supporting other parts of the product, so it's simpler to identify the monetization models that make sense to adopt. Keep reading to learn the difference between the freemium, tiered, and **pay-as-you-go (PAYG)** monetization models.

The freemium model

The freemium monetization model is the easiest one to understand and put into practice. Jarid Lukin and Fred Wilson first used the term "*freemium*" in 2006. Wilson is a prolific venture capitalist who published an article describing a business model where you give your base product away for free and then offer premium features that customers must pay for. He asked his readers for suggestions of names for that business model, and Lukin proposed the term "freemium". The name stuck, and it's been widely used since then. There's a subtle difference between the freemium model and a product trial. A trial is usually associated with a period, while freemium is about limiting the offer to a set of features.

You choose one or more features that you make available for free. The opportunity to have access to certain features makes it easy for customers to see the API working, even though in a limited fashion. Customers can test the API without paying a subscription and see how the free features can be useful. LanguageTool, a popular grammar-checking API, uses the freemium model to let customers use the service for free within limitations such as the number of characters per request. As the API producer, LanguageTool has the chance to show customers what the paid features look like and what their benefits are. The idea is that as customers use the free features often, they get engaged with the API and will eventually want to use all the features. And to have access to all the API features, customers will have to pay a monthly fee. The paid subscription can include all the features, or you can decide to offer various subscription tiers.

Tiered model

When you can identify different cohorts of users, you can package various features that each cohort will find useful. The tiered monetization model helps you create a subscription price for each set of features. The tiered model gained popularity in the early 2000s with **Software-as-a-Service**, or **SaaS**, companies such as Salesforce. At that time, companies started to offer different pricing tiers to different types of customers. The goal is to be able to provide more value to customers by making available the features that they care about the most. The tiered model has the advantage that it can be easy to upsell customers to higher tiers.

You evolve from a freemium model as the one you've read about before into two or more tiers with an increasing number of offered features or usage volume. Each tier includes all the benefits of the previous tiers plus added features or a higher volume of API requests. Typically, each tier is associated with a name representing each customer cohort. You can find tier names such as standard, professional, and enterprise. Naming tiers helps customers identify the right set of features that fit their needs. As with the freemium model, you have the opportunity to show customers features that are only available at a higher tier and entice them to upgrade. This type of monetization model works well with APIs where the operational cost is not associated with the number of requests being made.

PAYG model

If the cost to operate your API is directly associated with the number of requests users make, then the PAYG monetization model is your best choice. You can easily recognize this model from your electricity or water bills, where you only pay for what you consume. By making every API request billable, you ensure that your operational costs will never surpass your revenue. From the customer's perspective, the most significant advantage is that they only pay for what they consume. The barrier to using the API is relatively low because there are no flat fees as in the tiered model you've learned about before. **Amazon Web Services**, or **AWS**, uses this model effectively, making it very easy to start using any of the APIs it offers.

The PAYG model can also be used in conjunction with the freemium or tiered model. In the case of freemium, you can offer a number of API requests for free per month and charge for each request after the initial volume is reached. The advantage is that you still let users test your API but within limits that are operationally safe. You can use PAYG alongside the tiered model by providing the highest tier for customers that reach a high usage volume. After that volume is reached, customers start paying for each request. This approach makes it easy for larger companies to keep using your API with an easy-to-calculate expense. Whichever model you choose, pay attention to your own operational costs, which should cover not only the infrastructure to run the API but also ongoing customer support.

Support and documentation

The more popular an API is, the higher its support costs are. In the beginning, the only costs your API has are related to its development. During the API design and development stages, you spend time and resources ensuring you can launch your API. After that stage, you focus on having your API running, and you quickly start having infrastructure costs. Over time, as you start getting API users, significant costs will be related to support. More users interacting with your API means you'll have to provide more technical assistance. Even if your API looks like the easiest one to use, there will be users who will need help. The cost of support grows to the point where you introduce a new version that addresses most user issues. However, with any new API version introduced, more challenges will emerge, and more users will require support assistance.

Support is one of the most underrated parts of operating a successful API product. Without a viable support organization, you won't be able to keep your existing customers happy and evolve your API to attract new customers. There are many ways in which you can offer efficient support to your API users. I like to split support approaches into the categories of reactive support and proactive support. Reactive support includes any interaction that originates with users and evolves into an action to mitigate and eventually fix the problem being described. Some reactive support channels are email or chat, social media, and developer community forums. In general, the more public a support channel is, the higher the priority you should give to responding to requests. That is because you want to keep other potential users engaged by the quality of your interactions. Therefore, reactive support involves interaction with your employees and generates a real cost, which you can quantify mainly in the form of salaries. On the other hand, proactive support consists of any activity that you perform to identify challenges that API users are having. By anticipating the needs of your users, you avoid having to interact with them on a case-by-case basis. Instead, you provide them with enough information that they can fix issues by themselves without your intervention. Examples of proactive support include monitoring your API for problems, identifying challenging usage patterns, creating a training program, and providing comprehensive documentation. As you can see, support can become the highest cost you'll have to deal with during the lifetime of your API. While documentation alone can't completely eliminate the need for support, it can fulfill most of the user's needs. Documentation involves not only a complete API reference but also a developer portal, tutorials, and **frequently asked questions or FAQs**. Let's look at each of those types of documentation in detail.

An API reference is a document that describes in detail how the API works. The API reference has information about the API endpoints, parameters, responses, errors, authentication, and any other information needed to use the API. An API reference can also include usage examples and ways to start using each one of the available operations quickly. One such way is the ability to generate code snippets or complete **software development kits**, or **SDKs**, which can make it easy for users to interact with the API. While you can write an API reference from scratch, you often generate it from a machine-readable API definition document. You first create the API definition document during your API design stage and then convert it into a human-readable document that can be used as the API reference. While an API reference alone can help users in most situations, there are cases where having access to a developer portal is helpful.

API developer portals are living documents, typically websites, where users can find pointers to all existing information about the API. An example of a comprehensive API developer portal is the one that Notion, a widely used productivity service, put together. Notion's developer portal includes a "get started" section, the API reference, examples of what's possible to do with its API, a way to sign up as a partner, links to community forums, tutorials, FAQs, and a way to contact support. As you can see, the goal of Notion's developer portal is to provide an easy-to-find centralized resource that users can go to whenever they need assistance. An exciting part of the portal is its **Guides** section, which offers ready-to-follow tutorials for common use cases.

API tutorials are step-by-step guides that give API users instructions for using the API in specific scenarios. Each tutorial is directed at a specific use case that API users care about. The goal of API tutorials is to help API users learn how to perform specific workflows. Users don't have to discover all the steps by themselves. Instead, they can simply follow the tutorial and have all the information and code samples readily available. Tutorials can be made available in the form of written articles and also via videos or recorded webinars. Twilio, a popular communications platform including voice and text messages, has a vast amount of video tutorials available online. Its approach was to create short videos dedicated to single features. Users can either watch individual videos or opt to follow a series of related tutorials. The rule here is to find how your users like to consume information and make tutorials available using those channels.

While a comprehensive API reference and a good coverage of tutorials will help you create an engaging API **user experience** (**UX**), it can also help you reduce the cost of support. If, during support, you capture and categorize questions and their respective answers, you'll be able to translate those interactions into a list of FAQs. You can also analyze how often users access each FAQ over time. You can even make the FAQs list dynamic by showing first the questions with the highest number of visualizations. Having the list of FAQs handy whenever there's an incoming support request helps you quickly point users to the correct answer. To manage this process, you can use tools such as Zendesk, a well-known support service that enables you to manage the whole process. One thing to keep in mind whenever publishing answers to FAQs is the security implications of the information you're sharing. Keep reading to learn about the different aspects of API security.

Security

There's no way to fully protect an API from security threats. At least, that's what experts in security will tell you. In 2022, I talked with API security experts from different companies. I wanted to understand why there's a belief that all APIs are insecure. What I learned is that API security involves a combination of actions that help you reduce your risk but never eliminate it completely. The most important aspect of API security is knowing how your API is being used. Knowing whether you have any risks of being breached is an important first step even if you're not implementing any security measures. Let's dig into what that means.

Logging and monitoring

You can only improve what you can measure. If you aim to enhance your API security, then you need to start by measuring your API. There are two ways to gain knowledge about how your API runs. You can have qualitative information in the shape of events related to how users interact with your API, and you can also obtain quantitative information on how your API is performing. The conjunction of both types of information can help you understand what kinds of requests generate potential security threats. Logging lets you see each request as it was made by API users. Depending on the level of verbosity, you can see the type of request, the resource the request refers to, and even the full request and its corresponding response. Over time, you will accumulate information about all the API requests in a log. The log should be easy to search, and you should also be able to quickly identify usage patterns and their periods. The information you obtain from logs can be used in conjunction with monitoring data to identify threats and possible mitigations. You use API monitoring solutions to periodically check whether your system is behaving normally and, if not, determine which parts of the system need to be fixed. You can, for example, identify that your system is getting slow because you're receiving too many requests. In that case, you would look at solutions to limit the rate of requests that API users can make.

Rate-limiting

An increase in usage of your API often translates into more resources needed from the infrastructure that supports it. When usage is evenly spread across all API consumers, you can safely assume you need to increase your infrastructure's ability to meet higher demand. However, there are situations when you have peak usage or the request pattern of certain API consumers drains your systems. As you've seen before, those scenarios can be identified using a combination of logging and monitoring. Rate-limiting is a solution that you can employ to prevent API users from making requests at a frequency so high that your infrastructure wouldn't be able to handle them. Most API gateway products offer rate-limiting solutions, so you don't have to build anything from scratch. When configuring rate-limiting, it's vital to define the maximum number of requests per period and also to make users aware that they've reached that limit. A usual rule of thumb is to identify the maximum number of requests that your servers can handle per minute. Then, you divide that number by the total number of API users and get the limit per user. For example, if the maximum number of requests your infrastructure can handle per minute is 3,600 and you expect 100 simultaneous API consumers, your conservative rate limit would be 36 requests per minute. As a rule of thumb, you can start with a less conservative limit—say, 60 requests per minute or the equivalent of 1 request per second. Then, you can adapt the value over time by monitoring how your system reacts. One area particularly vulnerable to attacks involving a high frequency of requests is API authentication, so special attention is advised.

Authentication and authorization

There are cases where an API doesn't need to identify who its users are. The typical scenario is different. With most API products, you want to know who your users are so that you can use a monetization model such as the ones you've read about before. To identify users, you need to have a way to let them authenticate themselves for each and every request they make. Different authentication schemes can be used directly with existing API gateway solutions. Let's look at the most used ones and see how they can help you increase the security of your API. The first authentication scheme is called HTTP basic authentication. This is a standardized way of asking users to provide a username and password to access a web resource. Any HTTP client can use this authentication method, and it doesn't require any specific implementation to work. However, it's not the most secure because the credentials are not encrypted, and users need to send their usernames and passwords on every request. An evolution of this authentication scheme is the use of API keys. Instead of asking users to send their usernames and passwords, you let them create API keys that they send on every request. The advantage is that users can have many API keys that they can use in different scenarios and even revoke if needed. The disadvantage of API keys is that it's easy for API users to misuse them by sharing them with third parties. For those situations, you have another protocol, OAuth, which can be used for both authentication and authorization. With OAuth, you can effectively identify the third-party application that makes requests on behalf of a user. If you want to, you can restrict access to certain third parties without affecting what API users can do. But the strongest point of OAuth is its embedded authorization controls. With OAuth, you can define scopes, or groups of actions that users are allowed to do. Then, you can restrict which scopes a third-party application is allowed to perform on behalf of users. For increased security, all these checks can be done at the API gateway level even before they reach the API servers.

Summary

You now know how to increase the business value of your company by using an API monetization model. You know how to provide API support and use documentation to increase user satisfaction. You also understand the different factors that affect the security of your API product.

You started by identifying business value and understanding that different stakeholders perceive it differently. You learned how to analyze your API to understand the business value it can bring to a company. Then, you went deep into ways to generate revenue by looking at the freemium, tiered, and PAYG monetization models. You learned that you can use the monetization models individually or in conjunction to achieve maximum potential. Afterward, you got to know the different elements of API user support and documentation. You learned that support has the highest operational cost after an API is running and explored ways of reducing that cost. Among those are good documentation and FAQs. Finally, you explored the various attributes of API security. You started by seeing that the most important thing is measuring your API usage using logging and monitoring tools. After that, you learned about rate-limiting your API as a way to make sure that it's always running smoothly. To end the chapter, you saw the most important API authentication and authorization options.

Here's a list of the things that you learned during this chapter:

- The meaning of business value and its relation to various company stakeholders

- The ways APIs can increase a company's business value

- API monetization models that help you generate revenue

- How to use the freemium, tiered, and PAYG monetization models together or individually

- The cost of API support

- Ways to automate API support by generating an up-to-date list of FAQs and their answers

- The importance of having a high-quality API portal that includes a reference, tutorials, onboarding instructions, and community forums

- The various attributes of API security, including logging, monitoring, rate limits, authentication, and authorization

Thank you for following me on the journey of building an API product. Hopefully, you've enjoyed it as much as I have. At this point, you have the foundational elements of what an API product is. You should now be ready to learn about the process of building and maintaining API products. In the next chapter, you'll cover the stages of the API life cycle. Keep reading to learn about the design, implementation, release, and maintenance life cycle stages.

4
API Life Cycle

The life of an API product consists of a series of stages. Those stages form a cycle that starts with the initial conception of the API product and ends with the retirement of the API. The name of this sequence of stages is called a life cycle. This term started to gain popularity in software and product development in the 1980s. It's used as a common framework to align the different participants during the life of a software application or product. Each stage of the API life cycle has specific goals, deliverables, and activities that must be completed before advancing to the next stage. There are many variations on the concept of API life cycles. I use my own version to simplify learning and focus on what is essential. Over the years, I have distilled the API life cycle into four easy-to-understand stages. They are the design, implementation, release, and maintenance stages. Keep reading to gain an overview of what each of the stages looks like.

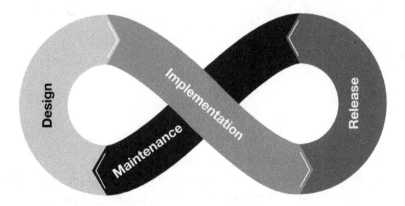

Figure 4.1 – The API life cycle

The goal of this chapter is to provide you with a global overview of what an API life cycle is. You will see each one of the stages of the API life cycle as a transition and not simply an isolated step. You will first learn about the design stage and understand how it's foundational to the success of an API product. Then, you'll continue on to the implementation stage, where you'll learn that a big part of an API server can be generated. After that, the chapter explores the release stage, where you'll learn the importance of finding the right distribution model. Finally, you'll understand the importance of versioning and sunsetting your API in the maintenance stage.

After reading the chapter, you will understand and be able to recognize the API life cycle's different stages. You will understand how each API life cycle stage connects to the others. You will also know the participants and stakeholders of each stage of the API life cycle. Finally, you will know the most critical aspects of each stage of the API life cycle.

In this chapter, you'll learn about the four stages of the API life cycle:

- Design
- Implement
- Release
- Maintain

Design

The first stage of the API life cycle is where you decide what you will build. You can view the design stage as a series of steps where your view of what your API will become gets more refined and validated. At the end of the design stage, you will be able to confidently implement your API, knowing that it's aligned with the needs of your business and your customers. The steps I take in the design stage are as follows:

- Ideation
- Strategy
- Definition
- Validation
- Specification

These steps help me advance in holistically designing the API, involving as many different stakeholders as possible so I get a complete alignment. I usually start with a rough idea of what the ideal API would look like. Then I start asking different stakeholders as many questions as possible to understand whether my initial assumptions were correct.

Something I always ask is why an API should be built. Even though it looks like a simple question, its answer can reveal the real intentions behind building the API. Also, the answer is different depending on whom you ask the question. Your job is to synthesize the information you gather and document pieces of evidence that back up the decisions you make about the API design. You will, at this stage, interview as many stakeholders as possible. They can include potential API users, engineers who work with you, and your company's leadership team. The goal is to find out why you're building the API and to document it. Once you know why you're building the API, you'll learn what the API will look like to fit the needs of potential users.

To learn what API users need, identify the personas you want to serve and then put yourself in their shoes. You've already seen a few proto-personas in *Chapter 2*. In this API life cycle stage, you draw from those generic personas and identify your API users. You then contact people representing your API user personas and interview them. During the interviews, you should understand their JTBDs, the challenges they face during their work, and the tools they use. From the information you obtain, you can infer the benefits they would get from the API you're building and how they would use the API. This last piece of information is critical because it lets you define the architectural style of the API.

By knowing what tools your user personas use daily, you can make an informed decision about the architectural style of your API. Architectural styles are how you identify the technology and type of communication that the API will use. For example, **REST** is one architectural style that lets API consumers interact with remote resources by executing one of the HTTP verbs. Among those verbs, there's one that's natively supported by web browsers—**HTTP GET**. So, if you identify that a user persona wants to use a web browser to consume your API, then you will want to follow the REST architectural style and limit it to HTTP GET. Otherwise, that user persona won't be able to use your API directly from their tool of choice. Something else you'll want to define is the capabilities your API will offer users. Defining capabilities is an exercise that combines the information you gathered from interviews. You translate JTBDs, benefits, and behaviors into a set of capabilities that your API will have. Ideally, those capabilities will cover all the needs of the users whom you interviewed. However, you might want to prioritize the capabilities according to their degree of urgency and the cost of implementation. In any case, you want to validate your assumptions before investing in actually implementing the API.

Validation of your API design happens first at a high level, and after a positive review, you attempt a low-level validation. High-level validation involves sharing the definition of the architectural style and capabilities that you have created with the API stakeholders. You present your findings to the stakeholders, explain how you came up with the definitions, and then ask for their review. Sometimes the feedback will make you question your assumptions, and you must refine your definitions. Eventually, you will get to a point where the stakeholders are all aligned with what you think the API should be. At that point, you're ready to attempt a low-level validation. The difference between a high-level and a low-level validation is the amount of detail you share with your stakeholders and how technical the feedback you expect needs to be.

While in high-level validation, you mostly expect an opinion about the design of the API, in low-level validation, you actually want the stakeholders to test the API before you start building it. You do that by creating what is called an **API mock server**. It allows anyone to make real API requests to a server as if they were making requests to the real API. The mock server responds with data that is not real but has the same shape that the responses of the real API would have. Stakeholders can then test making requests to the mock server from their tools of choice to see how the API would work. You might need to make changes during this low-level validation process until the stakeholders are comfortable with how your API will work. After that, you're ready to translate the API design into a machine-readable definition document that will be used during the implementation stage of the API life cycle. The type of machine-readable definition depends on the architectural style identified earlier. If, for example, the architectural style is REST, then you'll create an **OpenAPI** document. Otherwise, you will work with the type of machine-readable definition most appropriate for the architectural style of the API. Once you have a machine-readable API definition, you're ready to advance to the implementation stage of the API life cycle.

Implementation

Having a machine-readable API definition is halfway to getting an entire API server up and running. I won't focus on any particular architectural style, so you can keep all options open at this point. The goal of the machine-readable definition is to make it easy to generate server code and configuration and give your API consumers a simple way to interact with your API. Some API server solutions require almost no coding as long as you have a machine-readable definition. One type of coding you'll need to do—or ask an engineer to do—is the code responsible for the business logic behind each API capability. While the API itself can be almost entirely generated, the logic behind each capability must be programmed and linked to the API. Usually, you'll start with a first version of your API server that can run locally and will be used to iteratively implement all the business logic behind each of the capabilities. Later, you'll make your API server publicly available to your API consumers. When I say publicly available, I mean that your API consumers should be able to securely make requests.

One of the elements of security that you should think about is authentication. Many APIs are fully open to the public without requiring any type of authentication. However, when building an API product, you want to identify who your users are. Monetization is only possible if you know who is making requests to your API. Other security factors to consider have already been covered in *Chapter 3*. They include things such as logging, monitoring, and rate limiting. In any case, you should always test your API thoroughly during the implementation stage to make sure that everything is working according to plan.

One type of test that is particularly useful at this stage is contract testing. This type of test aims to verify whether the API responses include the expected information in the expected format. The word *contract* is used to describe the API definition as something that both you—the API producers—and your consumers agree to. By performing a contract test, you'll verify whether the implementation of the API has been done according to what has been designed and defined in the machine-readable document. For example, you can verify whether a particular capability is responding with the type of data that you defined. Before deploying your API to production, though, you want to be more thorough with your testing. Other types of tests that are well suited to be performed at this stage are functional and performance testing. Functional tests, in particular, can help you identify areas of the API that are not behaving as functionally as intended. Testing different elements of your API helps you increase its quality. Nevertheless, there's another activity that focuses on API quality and relies on tests to obtain insights.

Quality assurance, or QA, is one type of activity where you test your API capabilities using different inputs and check whether the responses are the expected ones. QA can be performed manually or automatically by following a programmable script. Performing API QA has the advantage of improving the quality of your API, its overall user experience, and even the security of the product. Since a QA process can identify defects early on during the implementation stage of an API product, it can reduce the cost of fixing those defects if they're found when consumers are already using the API. While contract and functional tests provide information on how an API works, QA offers a broader perspective on how consumers experience the API. A QA process can be a part of the release process of your API and can determine whether the proposed changes have production quality.

Release

In software development, you can say that a release happens whenever you make your software available to users. Different release environments target different kinds of users. You can have a development environment that is mostly used to share your software with other developers and to make testing easy. There can also be a staging environment where the software is available to a broader audience, and QA testing can happen. Finally, there is a production environment where the software is made available generally to your customers. Releasing software—and API products—can be done manually or automatically. While manual releases work well for small projects, things can get more complicated if you have a large code base and a growing team working on the project. In those situations, you want to automate the release as much as possible with something called a **build process**.

During implementation, you focus on developing your API and ensuring you have all tests in place. If those tests are all fully automated, you can make them run every time you try to release your API. Each build process can automatically run a series of steps, including packaging the software, making it available on a mock server, and running tests. If any of the build steps fail, you can consider that the whole build process failed, and the API isn't released. If the build process succeeds, you have a packaged API ready to be deployed into your environment of choice. Deploying the API means it will become available to any users with access to the environment where you're doing the release. You can either manage the deployment process yourself, including the servers where your API will run, or use one of the many available API gateway products. Either way, you'll want to have a layer of control between your users and your API.

If controlling how users interact with your API is important, knowing how your API is behaving is also fundamental. If you know how your API behaves, you can understand whether its behavior is affecting your users' experience. By anticipating how users can be negatively affected, you can proactively take measures and improve the quality of your API. Using an API monitor lets you periodically receive information about the behavior and quality of your API. You can understand whether any part of your API is not working as expected by using a solution such as a **Postman Monitor**. Different solutions let you gather information about API availability, response times, and error rates. If you want to go deeper and understand how the API server is performing, you can also use an **Application Performance Monitor (APM)**. Services such as **New Relic** give you information about the performance and error rate of the server and the code that is running your API.

Another area that you want to pay attention to during the release stage of the API life cycle is documentation. While you can have an API reference automatically built from your machine-readable definition, you'll want to pay attention to other aspects of documentation. As you've seen in *Chapter 2*, good API documentation is fundamental to obtaining a good user experience. In *Chapter 3*, you learned how documentation can enhance support and help users get answers to their questions when interacting with your API. Documentation also involves tutorials covering the JTBDs of the API user personas and clearly showing how consumers can interact with each API feature.

To promote the whole API and the features you're releasing, you can make an announcement to your customers and the community. Announcing a release is a good idea because it raises the general public's awareness and helps users understand what has changed since the last release. Depending on the size of your company, your available marketing budget, and the importance of the release, you choose the media where you make the announcement. You could simply share the news on your blog, or go all the way and promote the new version of your API with a marketing campaign. Your goal is always to reach the existing users of your API and to make the news available to other potential users.

Sharing news about your release is a way to increase the reach of your API. Another way is to distribute your API reference in existing API marketplaces that already have their own audience. Online marketplaces let you list your API so potential users can find it and start using it. There are vertical marketplaces that focus on specific sectors, such as healthcare or education. Other marketplaces are more generic and let you list any API. The elements you make available are usually your API reference, documentation, and pointers on signing up and starting to use the API. You can pick as many marketplaces as you like. Keep in mind that some of the existing solutions charge you for listing your API, so measure each marketplace as a distribution channel. You can measure how many users sign up and use your API across the marketplaces where your API is listed. Over time, you'll understand which marketplaces aren't worth keeping, and you can remove your API from those. This measurement is part of API analytics, one of the activities of the maintenance stage of the API life cycle. Keep reading to learn more about it.

Maintenance

You're now in the last stage of the API life cycle. This is the stage where you make sure that your API is continuously running without disturbances. Of all the activities at this stage, the one where you'll spend the most time will be analyzing how users interact with your API. Analytics is where you understand who your users are, what they're doing, whether they're being successful, and if not, how you can help them succeed. The information you gather will help you identify features that you should keep, the ones that you should improve, and the ones that you should shut down. But analytics is not limited to usage. You can also obtain performance, security, and even business metrics. For example, with analytics, you can identify the customers who interact with the top features of your API and understand how much revenue is being generated. That information can tell you whether the investment in those top features is paying off. You can also understand what errors are the most common and which customers are having the most difficulties. Being able to do that allows you to proactively fix problems before users get in touch with your support team.

Something to keep in mind is that there will be times when users will have difficulties working with your API. The issues can be related to your API server being slow or not working at all. There can be problems related to connectivity between some users and your API. Alternatively, individual users can have issues that only affect them. All these situations usually lead to customers contacting your support team. Having a support system in place is important because it increases the satisfaction of your users and their trust in your product. Without support, users will feel lost when they have difficulties. Worse, they'll share their problems publicly without you having a chance to help. One situation where support is particularly requested is when you need to release a new version of your API.

Versioning happens whenever you introduce new features, fix existing ones, or deprecate some part of your API. Having a version helps your users know what they should expect when interacting with your API. Versioning also enables you to communicate and identify those changes in different categories. You can have minor bug fixes, new features, or breaking changes. All those can affect how customers use your API, and communicating them is essential to maintaining a good experience. Another aspect of versioning is the ability to keep several versions running. As the API producer, running more than one version can be helpful but can increase your costs. The advantage of having at least two versions is that you can roll back to the previous version if the current one is having issues. This is often considered a good practice.

Knowing when to end the life of your entire API or some of its features is a simple task, especially when there are customers using your API regularly. First of all, it's essential that you have a communication plan so your customers know in advance when your API will stop working. Things to mention in the communication plan include a timeline of the shutdown and any alternative options, if available, even from a competitor of yours. A second aspect to account for is ensuring the API sunset is done according to existing laws and regulations. Other elements include handling the retention of data processed or generated by usage of the API and continuing to monitor accesses to the API even after you shut it down.

Summary

At this point, you know how to identify the different stages of the API life cycle and how they're all interconnected. You also understand which stakeholders participate at each stage of the API life cycle. You can describe the most important elements of each stage of the API life cycle and know why they must be considered to build a successful API product.

You first learned about my simplified version of the API life cycle and its four stages. You then went into each of them, starting with the design stage. You learned how designing an API can affect its success. You understood the connection between user personas, their attributes, and the architectural type of the API that you're building. After that, you got to know what high and low-level design validations are and how they can help you reach a product-market fit. You then learned that having a machine-readable definition enables you to document your API but is also a shortcut to implementing its server and infrastructure. Afterward, you learned about contract testing and QA and how they connect to the implementation and release stages. You acquired knowledge about the different release environments and learned how they're used. You knew about distribution and API marketplaces and how to measure API usage and performance. Finally, you learned how to version and eventually shut down your API.

These are the topics that you covered in this chapter:

- The four stages of the API life cycle: design, implementation, release, and maintenance
- How the design stage plays a fundamental role in the success of your API product
- The connection between user personas, their attributes, and the architectural style of your API
- What to expect from high- and low-level design validations
- The value of having a machine-readable API definition in the implementation stage
- Ways to automate the implementation of most of your API infrastructure
- Contract tests and QA, which are fundamental to a healthy API where users have a positive experience
- The different types of release environments and how they're used
- The role of distribution in attracting and retaining API users
- The different kinds of API analytics and what information you can get from each
- How to do API versioning correctly, ensuring that users are well informed
- How to plan the shutdown of a feature or the whole API and communicate it to your user base in advance

I hope you liked reading this chapter and learned the fundamentals of the API life cycle. This chapter was an introduction to a more in-depth analysis of each of the four stages of the API life cycle. With the information you now have, you'll be able to understand in detail all the steps to build an API product. You'll start by focusing on the design stage of the API life cycle in the next chapter. Let's continue our journey of building an API product.

Part 2:
Designing an API Product

This part is a comprehensive exploration of API product design, covering key stages such as ideation, strategy, definition, validation, and specification. It provides an in-depth analysis of the strategy stage, emphasizing stakeholder identification, business objectives, and understanding user personas. Additionally, this part guides you through defining and validating API design using techniques such as mocks, UI integration, and stakeholder iteration. It concludes with guidance on selecting an API architectural type based on behaviors, refining definitions, and creating machine-readable representations with governance rules.

In this part, you'll find the following chapters:

- *Chapter 5, Elements of API Product Design*
- *Chapter 6, Identifying an API Strategy*
- *Chapter 7, Defining and Validating an API Design*
- *Chapter 8, Specifying an API*

5

Elements of API Product Design

Building an API product starts with designing what it will look like and how it will behave. To design an API product, you can follow a series of steps that help you align your ideas with what stakeholders need. This chapter will guide you through the five steps of API design. You will get to know how to connect the dots between API ideation, strategy, definition, validation, and specification. This chapter acts as an introduction to each of these steps, all of which will be expanded on later in this book.

This chapter begins by explaining the ideation step and how to execute it. You'll learn how to obtain a high-level vision of the API product and evolve it using techniques such as brainstorming. Then, you'll see what a good strategy looks like and how it can help you during other stages of the API life cycle. After that, you'll learn how to translate the ideas for your API product into a definition of its architectural style and capabilities. You'll validate those two elements by obtaining feedback from stakeholders. Finally, once your API definition has been validated, you'll learn how to use the OpenAPI and AsyncAPI specifications to create machine-readable definitions.

By the end of this chapter, you'll be able to understand and describe the design stage of the API life cycle. You'll know what the five steps of the design stage are and how they're connected. Finally, you'll see the importance of a machine-readable definition and how it can help you in other stages of the API life cycle.

In this chapter, you'll learn about the following API design topics:

- Ideation
- Strategy
- Definition
- Validation
- Specification

These topics constitute the API design process. While ideation can be understood as something that happens before the API design begins, we'll cover it in this chapter:

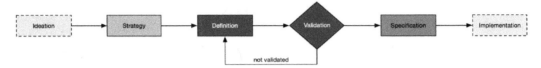

Figure 5.1 – The API design process

Technical requirements

This chapter introduces the OpenAPI and AsyncAPI specifications via **YAML Ain't Markup Language** (**YAML**) format. Even though you don't need any knowledge of these, you should have a minimum understanding of how configuration files work. In particular, knowing about configuration labels, attributes, and objects will be helpful.

Ideation

Every API product starts with an idea. Ideas can come from different stakeholders in different parts of an organization. Ideation is a process that helps organizations generate ideas for new or existing products and features. The ideation process is widely accepted among product design practitioners. The designers at IDEO, a design consultancy company, were the ones that started to use ideation in the 1990s. Ideation is a process that contains a series of steps that take you from not knowing how to solve a problem to having a clear direction for a solution. Let's look at those steps in detail.

You begin an ideation process by defining the problem that you're trying to solve. In the case of an API product, you could describe the problem as a high-level idea of what the API would solve. For example, for a company such as Stripe, the problem would be related to the difficulty that developers have in managing payments in their applications. In 2022, Stripe described itself as a service offering "payments infrastructure for the internet." This business statement is behind the problems that the company aims to solve. Likewise, the problem that you're solving should be aligned with the business objectives of your company. Otherwise, you might end up trying to solve the wrong problems. The problem statement should mention what the challenge is, as well as its scope. Without a clear scope, identifying possible solutions becomes hard. To succeed, you should get as much information as possible about the problem that you're trying to solve.

Obtaining information is something that you should do immediately after you've identified the problem and validated its alignment with business objectives. The goal at this point is to gather evidence that the problem exists by analyzing different sources of information. Then, you distill the problem into a definition that includes the evidence, the scope, and the business alignment. One of the sources of information is your list of customers or prospects if you're just starting a business. Interview people that you think have the problem you're defining. Understand how those people deal with the problem and what they do to solve it. With enough interviews, you'll be able to categorize the problem into different scopes. Another area of insight comes from analyzing market trends. Identify how the problem you're defining has been handled by the market in the past. You'll understand how the problem has evolved and which companies have offered solutions. From those companies, identify which ones are your potential competitors and analyze how they frame the problem. All this information will be valuable when you reach the point where you want to identify solutions to the problem.

Before identifying the solutions to the problem, you need to set the goals for the ideation process. Setting goals is important so that you don't get lost in the process and you know how to measure its success. One of the goals to identify is the criteria for evaluating the ideas. Here, it would be best if you understood the factors that influence the feasibility of the generated solutions. In some situations, what matters is the speed of delivery, while other situations favor solutions that are less expensive to implement. Your business objectives can guide you in identifying your idea evaluation criteria. Something else to define at this point is the number and diversity of ideas that you expect to obtain from the ideation process. The number of participants and the size of the problem scope are two factors that influence how ideas are generated, so pay attention to those things. Finally, at this stage, you must also identify any constraints or limitations that might shape the identified solutions. With your list of goals in hand, it's time to generate ideas.

Brainstorming is a technique widely used for generating ideas. The technique was created by Alex Osborn, co-founder of the advertising agency BBDO, in the 1940s. In one of his books, *Applied Imagination – Principles and Procedures of Creative Writing*, Osborn describes brainstorming as a way to use the brain to "storm a creative problem." Brainstorming encourages all participants to be creative and share their ideas openly without fear of being criticized. This technique is a great option for generating ideas for possible solutions to the problem that you identified earlier. The goal here is to let participants share as many ideas as they want. At this point, you're not going to categorize the ideas or judge their value.

Once you're done with the idea generation brainstorming, you can organize the ideas that the group came up with. You can ask brainstorming participants to describe their ideas individually. Then, you can have all participants vote on a category or theme for each idea. You can categorize ideas by relevance, business value, feasibility, cost, or any other criteria that you feel are relevant to your API product. Once you have split all the ideas into categories, it's time to prioritize them. Collaboratively, you and the participants can score each idea so that you can then find the most promising ones.

The list of the most promising ideas is valuable because it represents what the group of participants sees as the most viable solutions to your original problem. From this list, you can now conceptualize solutions. Defining concepts is an exercise that starts by creating mockups of the whole solution, including the part that is delivered by your API product. Mockups can go from simple sketches to usable prototypes. The final goal is to have something that represents each idea and lets stakeholders evaluate it. Depending on who your stakeholders are and the nature of the problem, each concept can be restricted to the API solution, or you might need to create a mockup of the API being used together with common tools used by potential users. Whatever you do, remember that your objective is to put the concepts in the hands of users and stakeholders so that they can give you feedback. Keep the concepts lean and to the point so that you don't spend too much time on concepts that you might have to throw away.

With the feedback you receive from users and stakeholders, you can understand which of the promising ideas are likely to have a satisfied user base. Whenever you receive negative feedback, use that information to refine the concept so that it approaches what stakeholders are looking for. Even though this part of the process can take time, you'll learn a lot from the interactions with stakeholders. You'll eventually get to a point where you'll know that you've reached several concepts that look like good solutions. To pick the one solution that you'll use as the foundation for your API product, you'll have to apply your own business and technical constraints. Apply the criteria that you used before to evaluate ideas, but this time, to this restricted set of concepts. You will reach a point where you'll have one single solution that best fits the needs of users and stakeholders and is also the most feasible to implement and the one that provides the most business value. This solution is the answer to what API product you're building.

The importance of following this ideation process lies in the way you involved potential users and internal and external stakeholders. You could come up with the idea for the API product yourself. However, you wouldn't have thorough validation and clear evidence of the need for the solution that you're proposing. The ideation process is the first step of API design because it lays the foundation for all the coming stages. From now on, everything you do will be based on the single solution that you identified during the ideation process. Next, you'll learn about the strategy stage of API design and how you can use it to further enrich the information about your solution while validating it with feedback from stakeholders.

Strategy

A good strategy starts with having a goal. However, it's much more than that. Good strategies define what you want to build, address business opportunities, align to reach a goal, provide evidence of business value, and identify implementation challenges. Let's start by identifying the API product that you'll build. Look back at the result of the ideation process. The solution that you have identified to be the best is exactly what you're going to build.

The second thing you want to do is define why you want to build the API product. Here, you have to gather as much evidence as you can to support the business value of the API you're designing. By now, you understand why the solution that you've identified is valuable. Let's rewind to the beginning of this chapter and dive deeper into identifying why the problem is valuable. From the user interviews, you should know that the problem affects a reasonable number of users. Now, let's map the problem and those potential users to value that is tangible to the business. To do that, go back to the interviews and extract the impact that the problem has on the lives of users. You can identify the impact by learning about whether users have to spend time or money because they have the problem. This is what I call the cost of the problem. In other words, having the problem has a cost to users, and they would prefer to have it fixed. The cost of the problem is one of the indicators of what price your solution should have. Other indicators include how much competitors are charging for a similar solution, if there are competitors, and how much the user cohort that you interviewed can spend to fix the problem. With all this information, you should be able to identify a price range for your API product. To calculate the business value, you must translate the price range into annual revenue for the size of the user cohort. By doing this, you will know how much revenue you'll be able to generate from the API product. And that is the evidence that can answer the question of why you should build the API product. One of the benefits of building the API is generating revenue, but there can be other outcomes.

Among the outcomes that an API product can generate, there's the ability to create an integrated product. Users often employ different solutions to overcome the challenges they have. An integrated product is a concept that I developed in 2017. The concept builds on the ideas of the whole product and augmented product. A whole product consists of all the features from different products that customers use to solve their needs. This idea was created by Regis McKenna, a marketer known for helping launch the first Intel microprocessor and Apple's first personal computer. The notion of the whole product was popularized by Geoffrey Moore in the book *Crossing the Chasm*. Augmenting a product is the act of making a product whole by adding all the features that users need to it. The concept of augmented products was created by Philip Kotler, a marketer and creator of the Total Product Concept strategy. The integrated product is a version of the original product that's been augmented by integrating features from other products to make it a whole product. Integrating an API product is fairly easy because it's programmable and simple to connect to other APIs. Instead of augmenting a product by building features or packaging the product with added features from partners, you can integrate it with other products. The same thing happens with other products when they're integrated with your API. One of the best outcomes of your API product is when someone else integrates it with their product. Your API product becomes a part of someone else's product, and their marketing efforts will help you grow. But to get there, you still have to build the API.

Knowing how you can go from an idea to a complete running API product is one of the tasks you have during the strategy stage. Planning the development of an API product doesn't mean that you know every detail in advance. At this point, what matters is that you understand what resources you estimate you'll need during development and how much you think the project will take to complete. It's important to know, for example, whether you'll need to hire people or whether you'll need to reallocate people from other projects. Another factor to include is an estimate of the risk that the project will fail. These pieces of information can be used to prioritize the API product's implementation against other projects in the company competing for resources. After you have put all the information together and the project has a positive review, you're ready to start talking to potential users of the API product.

As you saw in *Chapters 2* and *4*, understanding who your API users are is fundamental to delivering a successful product. Now, you'll learn what information you can obtain from user persona research. To begin with, user persona research helps you get in touch with real people. While some of those people might never use your API product, some others will eventually become your customers. By interacting with real potential customers, you'll get valuable feedback. With the information that you obtain from the interviews, you can build a list of user personas. User personas are nothing more than fictional characters representing a specific type of person who would use your API. As you know by now, each persona has unique characteristics, such as JTBDs, behaviors, and challenges. Let's see how you can identify those characteristics in more detail.

JTBDs are one of the most important elements to gather from interviewing potential users. A JTBD represents not just what people are trying to accomplish but also has the potential to identify what features your API product should offer. JTBDs represent the needs and motivations of your potential customers. While a JTBD refers to a task that someone is trying to accomplish, a feature is a solution that accomplishes the job. As an example, a JTBD involving payments could be "paying for an online service quickly and easily." A feature that would provide a solution for that JTBD could be a simple-to-use web form that connects to a payments API. If understanding the tasks your potential users spend time with is important, knowing how they would use your API also helps you shape your product.

One key area of API user persona research is identifying the behaviors of potential users of your API. The goal is to know how a user would consume your API if it already existed. This exercise helps you understand the attributes that your API should have to fit with the way users think and work. For example, knowing these behaviors will help you understand whether users prefer to make a single request and receive a large amount of data or whether they would rather like to make multiple requests that retrieve small amounts of information. It also helps you identify what type of authentication and authorization users prefer and even understand what content type users are more inclined to work with. Behaviors are also connected to user challenges, another area worth investigating.

While JTBDs represent the tasks that your potential users have to accomplish, challenges represent all the obstacles that users face while they're trying to complete their tasks. Knowing the challenges that users face helps you avoid directions that would create similar challenges. For example, a challenge related to online payments would be "not remembering a credit card number." By knowing about that specific challenge, you could avoid having a credit card number as input for your API. Challenges and JTBDs are tied together, and they're both part of how your potential users go through their daily activities. Another area to study is the benefits that users would get from using your API.

Knowing how your API makes people's lives easier helps you identify what capabilities you should implement. An API capability can correspond to a whole feature or be a part of it. Each benefit that users reveal can map directly to one capability. You work backward from the benefit to how your API would help users realize it. Following the previous examples related to payments, suppose a benefit is "easily making online payments without having to remember a credit card number." This benefit would map to a single capability of making a payment using a previously stored credit card. Another interesting thing about benefits is that you can use them in your marketing communication. In this case, you could announce your API by writing that "users will be able to easily make online payments without having to remember their credit card numbers." So, not only can user benefits inform you of what capabilities to implement but they can also guide you in your marketing activities. Another area that is critical in understanding how your API will be implemented is knowing what tools your potential users are already using in their daily tasks.

By identifying what tools your potential API consumers already use, you can understand what architectural style to implement. REST, GraphQL, gRPC, and SOAP are the most popular API architectural styles. Knowing which one to use is critical to the success of your API product. Each API architectural style has advantages and disadvantages, so understanding how to choose the right style is critical. First, let's see how the different architectural styles compare. First, you pick attributes such as discoverability, performance, and reliability. Then, you create a matrix that lets you compare architectural styles at a high level. It's important to remember that the attributes and their attribution to each architectural style are entirely up to you to define. As an example, let's see what an architectural style attribute matrix looks like:

	REST	GraphQL	gRPC	SOAP
Simple to use	✓	✓	✓	
Discoverable	✓			✓
Reliable	✓	✓	✓	
Performant	✓		✓	
Scalable	✓			
Observable	✓	✓	✓	✓

I follow two rules that help me decide which architectural style to use whenever I'm designing an API. First, I make sure that users can easily consume the chosen API architectural style with the tools they already use. There's no point in picking an architectural style that none of your potential API users can consume. Second, I guarantee that the chosen architectural style can work well with the operations and the type of data that the API will handle. By combining these two rules, I end up with an API that can be consumed by users while making it a good technical solution for handling the implemented capabilities.

By now, you should understand how the JTBDs, behaviors, challenges, benefits, and tools you use connect to your ability to identify the features, capabilities, and architectural style of your API. Next, you will learn how to define those characteristics.

Definition

After creating the strategy for your API, it's time to define how the API will behave. We can start defining the API architectural style by following the approach you read about before. To get started, let's describe the most popular API architectural styles and their advantages and disadvantages.

At the top of the popular API architectural styles is REST. As you learned in *Chapter 1*, this style is attached to the HTTP protocol and works well on the web in situations where API users want to both read and write information. However, it's worth noting that most implementations of REST are limited to reading data because using the HTTP GET method is less complex than using other methods that permit data manipulation. Other well-known architectural styles in 2023 include GraphQL and gRPC. Both architectural styles can easily be used on APIs that work across the internet. While GraphQL focuses on the ability to retrieve and manipulate arbitrary data, gRPC aims to let API consumers execute procedures remotely in a fast and efficient manner. Another architectural style to consider is SOAP, especially if you're working in environments that require compatibility with legacy systems. SOAP works well on different communication protocols, such as HTTP, **Simple Mail Transfer Protocol (SMTP)**, and **Java Messaging Service (JMS)**. SOAP is considered more difficult and resource-consuming than REST and other more recent architectures. Nevertheless, SOAP is still widely used inside enterprise networks. Given that there are many architectural styles, knowing how to pick the right one is key to designing an API that will meet the needs of users. Let's look at one way to match architectural styles with the expectations of API consumers.

A good way to understand what architectural styles users can consume is to create what I call a compatibility matrix. The rows of the matrix represent each of the tools used by potential API consumers, while each column represents one of the existing architectural styles. Then, you test each combination of tools and architectural style and document its compatibility. Let's see what such a matrix would look like:

	REST	GraphQL	gRPC	SOAP
Tool 1	✓	✓	✓	
Tool 2	✓			✓
Tool 3			✓	

As an example, following the payments scenario that I've been using, suppose that users tell you that they use Slack, a popular chat application. Your payment capability would have to make it easy for users to make a payment from a Slack message. Analyzing Slack's documentation, you find out that it follows the REST architectural style. Specifically, it makes an HTTP POST request with a particular payload to your API every time someone clicks on a button. So, now, you know that you wouldn't want to use GraphQL, gRPC, or SOAP as your API's architectural style. Otherwise, users wouldn't "be able to easily make online payments without having to remember their credit card numbers" from inside Slack – this sentence was derived from one of the API benefits that we went over previously in this chapter.

Finding the right fit between architectural styles and the tools that potential API consumers use is interesting. However, it's not enough to understand what the best architectural style for your API is. Something to consider is the number of potential API consumers for each of the tools listed. If your choice of architectural style can only be used by a tiny fraction of potential API consumers, then you may need to pick a different option. Or, you could test how your API would be used by choosing an architectural style that's easy to implement and consume. Another possibility is to decide what architectural style to implement based on the popularity of the most compatible tool. By following this approach, you can gain from making your API easy to integrate with tools that are widely used or in a growth stage. In any case, something else to consider is that you should identify the capabilities of your API and see how they can be used from the tool of choice.

Among the different options to identify what API capabilities to implement, one solution is to translate the JTBDs into what potential consumers would require. Since each JTBD consists of a description of something that users typically do to complete their tasks, this information can be used to identify the API's functional requirements. For example, if a JTBD is "paying for an online service quickly and easily," then a capability could be identified as "accept payments between users and online services." The way to describe capabilities is to think from the perspective of the API, not the consumer. You want users to understand what the API has to offer so that they can use the provided capabilities in their daily lives. Additionally, you want users to understand the benefits of using your API product.

Benefits are more powerful than capabilities because they help users visualize what they can get from using the API. The connection between capabilities and benefits is something that users need to be able to understand so that they learn what your API has to offer. Mapping users' benefits helps you work backward from what users would value to identifying what you should offer them. By following this way of working, your chances of aligning your API capabilities with users' needs increase. While some capabilities link to single benefits, others can address several benefits at once. You can make a map of how each capability addresses one or more benefits, and you can use those connections to prioritize your roadmap. Something else that can influence the priority of your implementation is the number of constraints that are placed on your API product.

While the architectural style and the API capabilities are things that users should be aware of, the underlying constraints should be invisible to consumers. Whoever uses your API product shouldn't feel any friction related to whatever constraints exist. And they can take many forms and come from different places. So, you must learn how to identify constraints and how to mitigate them. Among the different types of constraints, the three most important types are the ones related to your business, the ones that come from the choice of the architectural style, and the ones related to challenges shared by the personas. Business constraints exist because of a lack of full alignment between what the business needs and what the API product promises to offer. Most business constraints are related to costs and the promise of generating revenue. Along the way, while implementing your API, you might be asked to invest fewer resources in certain areas so that the cost doesn't rise. Or, you might be asked to let go of certain parts of the API product, such as documentation or support, to decrease the chances of incurring high costs over time. As the owner of the API product, one of your jobs is to find the right balance between what is possible and what is needed while following any existing constraints.

Business constraints are the ones that have the highest influence on the success of your API. That's because, by definition, an API product needs to provide value to your business. If there's no value to provide, then there's no case for building an API product. To generate value, you need to find what the business needs and provide that. As you saw in *Chapter 3*, value doesn't always translate into revenue. However, it must be aligned with the vision of the business, including whatever shortcomings and constraints it creates. An example of a business constraint is the definition of how the API can be accessed. While some APIs are built to be freely accessed, others are supposed to only be available behind a paid subscription. Paid APIs require you to implement a way to measure usage and restrict access in situations where consumers aren't allowed to make requests. That, by itself, is considered a constraint. Overall, business constraints include rate limiting, authentication, access control, usage restrictions, data privacy, and **service-level agreements (SLAs)**. Some of the constraints are easier to overcome than others. Some are even directly addressed by API gateway solutions. Kong, a popular API gateway, lets you configure rate limits using one of their plugins. Azure, another API gateway solution, gives you the ability to configure authentication out of the box. The second type of constraint that you read about before is related to the architectural style that you choose for your API. Depending on what foundation you use to build your API, you'll have to follow different rules, and you'll have different constraints. Examples of architectural style constraints include being restricted to using a specific communication protocol, such as HTTP, having a limit on the size of the payload used to send information from the API to the consumer, and making all information retrieval operations cacheable. Each architectural style documents a list of constraints and provides guidance on how you should implement your API so that you can embrace them. Finally, there are constraints that you can only understand once you start interacting with potential API consumers. While there are users who will happily adapt the way they work to use your API, what most of them want is precisely the opposite. Most users want your API product to work with the processes and tools they're already accustomed to. Constraints that involve personas are the hardest to solve. In most situations, those constraints require interoperability with other systems. Customers often feel that your API is merely one piece of their existing workflows. One way to prevent finding persona-related constraints too late is to involve consumers in the API design validation process. Keep reading to learn more about how API design validation works.

Validation

As introduced in *Chapter 4*, validation is one of the steps in the API design stage. Its goal is to confirm assumptions and make sure that the API product is designed according to what stakeholders expect. You engage in the validation process and make any necessary changes until you reach a point where you can confirm that your API design is solid. The idea is to involve stakeholders, including potential API consumers, so that they can give you feedback early on. You start with a high-level approach where you attempt to validate the architectural style and the API capabilities. Once you have positive feedback, you go deeper and perform a low-level validation where you gather information on how potential users interact with your API. Let's start by exploring the high-level validation approach.

The goal of high-level validation is to obtain feedback from stakeholders about abstract concepts that define how your API will behave. The feedback you get helps you understand whether you need to change the direction you're following while creating your API product. The architectural style is one of the concepts that you can validate at this stage. Even though you learned how to guarantee that your chosen architectural style is a good fit earlier, validating it with feedback from real people gives you more confidence. To validate an architectural style, ask potential consumers how difficult it is for them to interact with your API. That information will give you quantitative information that you can use to decide whether you need to perform individual interviews. If you have a reasonable percentage of stakeholders that find it hard to use your API, then you can select a sample group and do interviews to understand the reasons why. You might end up in a situation where stakeholders find it complicated to consume your API from within the tools they use, or they might tell you that they're not happy at all with those tools. Whatever the reasons, they will help you adjust your architectural style so that you can bring it closer to what your potential users would prefer. Once you get positive feedback from a reasonable number of stakeholders, you can move to the next validation step, where you'll confirm that your API product capabilities are aligned with the needs of potential users. You will follow an approach similar to the one you just learned about. However, this time, you'll want to understand how consumers rate each of the capabilities that you've designed. If you recall from earlier in this chapter, you can represent each capability by the benefit that it brings to users. At this stage, you can ask users how much each one of the identified benefits is important to them. Then, you can map each benefit to one or more capabilities and list them in order of importance, starting with the most important at the top. You'll soon understand which capabilities are must-haves and which ones aren't important to build. Having this information will help you prioritize the implementation of your API product and will save you costs and resources later in the API-building process. Another type of information that you can obtain is related to capabilities that you might have missed before. Make sure you have at least one open question asking potential users what capabilities they would add and why. During this step, you might end up finding surprising information that leads you to design new capabilities. After you have gone through both high-level validations, you're ready to understand the practical details of how users would consume your API and what difficulties they might encounter. To do that, let's look at how you can perform a low-level validation.

If, at a high level, you want to understand the abstract concepts that make your API product easy to use, at a low level, you want to know how your API would be used in a real-life scenario. The first thing you must do is find a way to let potential users interact with your API so that they can understand what it would feel like. However, you don't want to spend resources building an API at this point. Instead, you need to create a mockup of your API capabilities, available in the architectural style that you have chosen. To do that, you must create an API mock server using one of the many available tools. Postman, for instance, offers an easy way to create mock servers for a variety of API architectural styles. You can easily create a REST mock server where you expose the capabilities that you have previously validated. At this point, you don't want to go too deep into technical details because anything can still change, depending on the feedback you receive from stakeholders. You want to make the mock server available as quickly as possible and analyze how potential users are interacting with it. What matters at this point is that you obtain information on the number of requests that users make, the tools they use to make those requests, and the results of those requests. You want to find information such as the number of errors generated when users make requests using certain tools, or patterns on the requests that users make to understand if some capabilities should be isolated or grouped. Overall, your goal is to obtain usage details as if you had a real API working. Information about errors can be interesting, especially if the errors are related to users not providing information correctly when making requests. You can understand whether consumers are finding it complicated to use certain capabilities because they don't have all the required information, and you can adjust your approach depending on what you find. Even with all this information at your reach, you still have to engage with some of the users to see what they do with the information they get from the API mockup. You want to know whether the provided output is easy to use by consumers so that you can adjust its format and content if required.

As you've seen, API design validation is not a linear process. Instead, you participate in a feedback loop that helps you learn how potential users experience your API. With that information, you can make informed decisions to potentially change parts or the whole of the API design. Iteration after iteration, you sculpt your API, removing any imprecisions. The final result is a better API that fulfills the needs of your users. Once you're happy with what you have, you can move to the next step of API product design, where you translate the definition into something that can be used by machines. Continue reading to learn about API specifications and machine-readable definitions.

Specification

The API's specification is directly related to the chosen architectural style. In other words, there's at least one specification that's compatible with each architectural style. Sometimes, the specification is even a part of the architectural style. For example, a REST API can be defined using the OpenAPI specification. However, OpenAPI can also define APIs that don't follow the REST architectural style. As a different example, let's say you define a GraphQL API using the architectural style documentation. There's no need to use a separate specification. In either case, the outcome is that you will have a machine-readable API definition. This is important because API consumers communicate with APIs using tools that can read definitions. Without those machine-readable definitions, you would have to ask users to enter all the requested details by hand. Every time developers build an integration with an API, they can use a machine-readable definition to bootstrap the code that runs the integration. In this chapter, you'll learn how to create machine-readable definitions for synchronous and asynchronous APIs. Let's start with synchronous APIs using the OpenAPI specification format.

OpenAPI is a specification that can be used to define APIs that follow the REST architectural style. By using OpenAPI, you create a document that can be interpreted by software. To make that happen, the definition document has to be created using a format that a machine can interpret. OpenAPI definition documents can be written in both the YAML and JSON formats. YAML is a format that's widely used in technical documentation because it's easy for both humans and machines to understand. JSON, on the other hand, is more oriented at software interpretation and less friendly for you to understand. Some tools let you interactively create an OpenAPI definition. You can even convert the API mock server that you created previously into an OpenAPI definition. Nevertheless, let's take a look at a simple OpenAPI document so that you at least know what it looks like before you dig further into this book. The API that you'll see defined allows you to make online payments using a credit card. If you remember, this capability was identified earlier in this chapter. To make things easier to understand, let's break down the OpenAPI document into sections. In the first section, you'll find the identifier of the document's OpenAPI version and an object called `info`, which holds generic information about the API, followed by information on what servers can be used by consumers to make requests:

```
openapi: 3.0.3
info:
  title: Example payment API
  description: An example API for making online payments
    using a credit card.
```

```
    version: 1.0.0
 servers:
  - url: https://api.example.com/v1
```

As you can see, the OpenAPI version is 3.0.3. In the `info` object, you can find the API's title, a description of what it does, its version, and an example server definition.

The second section describes the HTTP paths and methods that the API is handling. This API follows the REST architectural style, so it lets users create payments by sending HTTP POST requests to the `/payments` path:

```
paths:
  /payments:
    post:
      summary: Make a payment
      description: Make an online payment using credit card
        information.
      requestBody:
        required: true
        content:
          application/json:
            schema:
              $ref: "#/components/schemas/PaymentRequest"
      responses:
        '201':
          description: Payment processed successfully
          content:
            application/json:
              schema:
                $ref:
                  "#/components/schemas/PaymentResponse"
        '400':
          description: Bad request
        '401':
          description: Unauthorized
        '403':
          description: Forbidden
        '500':
          description: Internal server error
```

The `paths` section is a bit longer than the first one. It not only defines the HTTP paths and methods but also identifies errors and what the request input and response look like. In this case, it identifies the `PaymentRequest` and `PaymentResponse` data types that will be defined in the components section. Now, let's see what's in the `components` section:

```
components:
  schemas:
    PaymentRequest:
      type: object
      required:
        - amount
        - cardNumber
        - cardExpiry
        - cardCvv
      properties:
        amount:
          type: number
          description: The amount to be charged to the
            credit card.
        cardNumber:
          type: string
          description: The credit card number.
        cardExpiry:
          type: string
          pattern: '^(0[1-9]|1[0-2])\/?([0-9]{4})$'
          description: The expiration date of the credit
            card in the format MM/YYYY.
        cardCvv:
          type: string
          description: The CVV number of the credit card.

    PaymentResponse:
      type: object
      properties:
        transactionId:
          type: string
          format: uuid
          description: The ID of the payment transaction.
```

As you can see, the `components` section defines both the `PaymentRequest` and `PaymentResponse` data types. You can see that the payment request is composed of the amount to be paid, the credit card number, the expiry date, and its **Card Verification Value (CVV)**. The payment response has an identifier for the payment transaction. It's easy to see that a payment request can take time to process, and thus it would be interesting to make the API asynchronous.

Let's follow the same process for defining the asynchronous version of the online payments API. As with OpenAPI, the AsyncAPI specification format lets you create a document using YAML. Let's break the definition into sections. At the top, you will find the identifier of the document's AsyncAPI version and the `info` object, which holds generic information about the API:

```
asyncapi: 2.0.0
info:
  title: Example asynchronous payment API
  version: 1.0.0
  description: An example asynchronous API for making
    online payments using a credit card.
```

This section is very similar to the one used in the OpenAPI definition. Here, the `servers` object has been split because it deserves more explanation:

```
servers:
  production:
    url: mqtt://api.example.com/v1
    protocol: mqtt
```

Unlike OpenAPI, the AsyncAPI specification lets you define the environment where your servers are operating and the protocol they're using. Here, you can see that the server that's been defined is in the production environment, and it's using the MQTT protocol. Let's continue with describing the available communication channels and operations:

```
channels:
  payments:
    publish:
      operationId: makePayment
      summary: Make a payment
      message:
        $ref: "#/components/schemas/PaymentRequestMessage"
    subscribe:
      operationId: paymentResult
      summary: Payment result
      message:
        $ref: "#/components/schemas/PaymentResponseMessage"
```

As you can see, there's a single channel called `payments`, which is enough for the online payments API that you're using as an example. Since this is an asynchronous API, you're defining two operations so that consumers can make payments and then obtain information about the status of a previous payment request. Consumers generate a payment transaction identifier and then issue a payment request. Upon terminating the payment operation, consumers will receive a notification that specifies the status of the transaction using the same identifier that was initially sent. To make that happen, you need to define the following messages and data types:

```
components:
  messages:
    PaymentRequestMessage:
      summary: Payment Request
      payload:
        $ref: "#/components/schemas/PaymentRequest"
    PaymentResponseMessage:
      summary: Payment Response
      payload:
        $ref: "#/components/schemas/PaymentResponse"

  schemas:
    PaymentRequest:
      type: object
      required:
        - transactionId
        - amount
        - cardNumber
        - cardExpiry
        - cardCvv
      properties:
        transactionId:
          type: string
          description: The ID of the payment transaction.
        amount:
          type: number
          description: The amount to be charged to the
            credit card.
        cardNumber:
          type: string
          description: The credit card number.
        cardExpiry:
          type: string
          description: The expiration date of the credit
            card in the format MM/YYYY.
        cardCvv:
          type: string
          description: The CVV number of the credit card.

    PaymentResponse:
      type: object
```

```
        properties:
          message:
            type: string
            description: A message indicating whether the
              payment was successful or not.
          transactionId:
            type: string
            description: The ID of the payment transaction.
```

As you've seen, the biggest difference between OpenAPI and AsyncAPI is that while the former is directed at the REST architectural style, the latter can be used for different styles and is appropriate for defining asynchronous APIs. Let's not focus on the implementation of these example APIs for now. Instead, the goal of the specification stage is to engage with whoever will take care of the implementation and work with them on defining all the details of the API. Having a machine-readable API definition has many advantages, including the ability to generate mock servers and production code, as well as bootstrap a developer portal where consumers can interact with the API.

Summary

By now, you understand the ideation, strategy, definition, validation, and specification API design steps. You know what each step is and how it influences the outcome of your API product. You also know how to execute the steps and connect them cohesively.

You began this chapter by learning how to expand a vision of an API product into actionable ideas. You learned how to use ideation techniques to involve stakeholders in designing your API. You also learned how to identify possible solutions from the ideas you generated. Following the ideation step, you learned how to define a strategy by identifying what you want to build, business opportunities, alignment to reach a goal, evidence of business value, and implementation challenges. You learned about the integrated product concept and how using it can position your API product for success. You also learned how JTBDs, behaviors, challenges, benefits, and tools can be used to define your API's features, capabilities, and architectural style. In the definition step, you understood how to identify the best architectural style for your API product. You then learned how to translate customer benefits into API capabilities. As part of the definition step, you also got to know what constraints are and how to identify them. Entering the validation step, you learned how to put the definition before stakeholders to obtain feedback. You learned about the difference between high- and low-level validation techniques. In particular, you learned that you can use an API mock server to obtain quantitative and qualitative data to help you refine your API design. Finally, near the end of this chapter, you learned about the role of API specifications. You understood that creating a machine-readable definition of your API makes different stages of the API life cycle easier to execute. You saw examples of machine-readable definitions by using the OpenAPI and AsyncAPI specifications with the YAML format.

These are some of the things that you learned in this chapter:

- How ideation helps organizations generate ideas for your API product
- The goal of brainstorming in the process of generating ideas
- How to evaluate, organize, and prioritize ideas to translate them into possible API product capabilities
- The elements of a good strategy
- The value of positioning your offering as an API that other products can easily integrate with
- How JTBDs, behaviors, challenges, benefits, and tools can be used to define your API's features, capabilities, and architectural style
- The advantages and disadvantages of the different API architectural styles
- How to translate customer benefits into API capabilities
- The different types of API constraints and how to use them in your favor
- The difference between high- and low-level API design validation
- How to use an API mock server to obtain qualitative and quantitative data to help you refine your API design
- Why having a machine-readable definition of your API makes executing different stages of the API life cycle easier
- How to implement practical examples of machine-readable definitions using the OpenAPI and AsyncAPI specifications

By now, you have all the information you need to dig deeper into different parts of designing your API product.

In the next chapter, you'll learn how to apply what you know to identify a winning strategy for your API. Keep reading to learn more.

6
Identifying an API Strategy

Behind every API product, there is a strategy. With it, you will find alignment with your business and your potential users. In this chapter, you will learn what a good strategy is and how to identify it. You will see how you can engage with stakeholders to keep the strategy aligned with business objectives. You'll also learn various techniques to define personas, their attributes, and their relation to different aspects of the API design.

You begin the chapter by learning what a strategy is. You'll learn its definition by seeing examples from well-known companies. Then, you'll get to know that goals, objectives, tactics, resources, and a timeline are all elements of a good strategy. You'll also see the difference between good and bad strategies. Continuing the chapter, you'll learn about the types of stakeholders, how you can involve them in decision-making processes, and how you can manage their relationship with the business. You'll then see the business objectives that the API will support and what costs it will generate. After that, you'll learn about user personas, how you can identify them and their attributes, and how they can aid you in aligning your API product. You'll see the relationship between personas, tools, and API architectural styles. Finally, you'll get to know how to identify user behaviors and how they're connected to the features that you'll offer.

By the end of the chapter, you'll know how to define a good strategy. You'll be able to connect stakeholders, user personas, and their behaviors to the success of your API product. Finally, you'll know how to employ techniques to identify your API architectural style and the features that you'll offer.

This chapter covers the following strategy topics:

- The meaning of strategy
- Stakeholders
- Business objectives
- Personas
- Behaviors

The meaning of strategy

The most strategic thing you can think of is choosing what not to do among all your options. At least that's what Michael Porter, one of the leading figures in the strategy world, would say. To the author of *Porter's Five Forces*, strategy is about understanding the trade-offs between what is available to do. If knowing what not to do is important, understanding how to get things done is critical. Another way to understand strategy is to view it as a cohesive set of actions that lead to meeting a goal. Take Toyota, a well-known automobile manufacturer. In the 1980s, Toyota decided to change the way they operated. They needed to improve production efficiency in the face of growing competition. To achieve that, they came up with the lean manufacturing approach, a methodology still followed today. Another example is what Netflix, a popular video streaming service, did when they realized their DVD rental business was dying. They had to reinvent themselves quickly, so in the early 2000s, they shifted their focus to streaming video online. If they hadn't done that, they would probably not exist today.

Sometimes a strategy is not pressed by competitors or diminishing markets. Instead, a long-term vision empowers companies to move forward along a pre-defined strategy. Tesla, for instance, is a company known for selling electric vehicles. They decided to differentiate themselves by creating a strategy to promote the concept of sustainable transportation. By doing that, Tesla has been building a solid brand and attracting loyal customers, leading to the growth of their business. Amazon had a similar goal of increasing its revenue. To do that, they defined a strategy to expand the company into new markets and product categories. Now, you can find Amazon available in almost any country, selling almost anything, not just books as it started with. Strategies can also be created to help companies increase their chances of being successful at achieving their goals. A good example is Apple's decision to shift focus to mobile devices in the early 2000s. The decision was, in fact, a strategy that has helped the company increase its market share since then. Not only was Apple able to grab a large portion of the mobile devices market, but it also gained a share selling personal computers.

The results of having a clear strategy can be pretty positive. While results help us understand how important a strategy is, its impact extrapolates its goals. For starters, having a clear strategy provides a direction that gets everyone aligned, and alignment ensures everyone is working to achieve the same goals—the ones identified in the strategy. It also provides an ever-present answer to anyone that wants to understand why they're moving in a specific direction. When everyone moves in the same direction, it becomes easy to focus on what matters. Anything that creates a detour can be seen as low priority and can be eliminated. By applying this exercise consistently, you can efficiently allocate resources that are aligned with the goals you want to achieve. A clear strategy is also a lever to the decision-making process. Decisions become consistent because there's a well-known path to be traveled. With this in mind, anyone in the organization can make decisions that put the company on the path and closer to the end goals. The strategy becomes a map that guides everyone in the same way. When everyone is moving in the same direction, it's easy to identify those whose contributions make the most significant strides. A sound strategy promotes accountability by applying metrics to identify contributors that put the company closer to its goals. It also enables clear communication channels that anyone can use to share their progress or ask for help when facing obstacles. Because there are obstacles in any path, the strategy acts as a north star, letting the organization adapt without losing track of its goals. To allow that, organizations can embrace a continuous improvement mindset. Through periodic evaluation and performance reviews, teams can adjust their direction and improve their velocity. Velocity is a function of speed and direction. Without a clear direction toward common goals, there is no velocity, only speed.

Goals are just one of the many components of a strategy. In fact, you could say that goals are the driver of a strategy. To be in the driver's seat, though, goals have to be clear and easy to communicate. They also have to provide alignment within the organization. Having goals that contradict how a company operates will put different teams on conflicting paths. While goals should be realistic, they only exist to determine the organization's intent at a high level. Objectives are, on the other hand, the realization of goals into specific targets. In the example of Apple you saw earlier, their goal was to obtain a steady yearly market share gain. Apple's objectives during that same period were related to the overarching goal but offered specificity. Apple wanted to expand its product line by developing innovative and user-friendly mobile devices that would appeal to a broad consumer base. They eventually built the iPhone among those new devices, attracting a horde of loyal customers. The strategy resulted in a growth in revenue from $5.7 billion in 2002 to almost $400 billion in 2022. To achieve results as consistently as Apple did, you need to find the right tactics to navigate your strategy. To begin with, strategy tactics need to be actionable. Tactics are the steps your team follows to get to the desired goals. Assigning tactics to specific teams is a way of making sure that they get completed. It also helps you to allocate the right resources at the right moment. You need to be able to deploy and withdraw resources whenever required along the way throughout your strategy. There are different strategy timelines to consider that give you different perspectives. Implementation timelines correspond to tactics and are a way to measure how long each group of actions takes to complete. Short-term timelines are associated with milestones that measure how far you are from reaching your desired goals. Finally, a long-term timeline is a way to zoom out and look at your strategy as a whole. All timelines should be monitored so that all stakeholders understand how far the organization is into the strategy.

But even before starting a strategy, it's a good idea to understand where you are and how far you are from reaching your desired goals. According to Richard Rumelt, author of the book *Good Strategy Bad Strategy*, having a clear picture of the current situation is the first attribute of a good strategy. Rumelt also advocates designing a strategy taking into account the strengths and weaknesses of your organization. By using those forces in your favor, your strategy will have a better chance of succeeding. Good strategies also focus on achieving specific and measurable results that will move the organization closer to its goals. This focus on results helps ensure that the organization is progressing and that resources are being used effectively. To understand the progress, you should have a feedback loop that measures results, external factors, and resource usage and adapts tactics for the whole strategy whenever necessary. By contrast, a bad strategy not only fails at addressing the initial challenges but is also built on a fragile foundation. Bad strategies can be identified because they often have vague goals, and their documents are filled with buzzwords and devoid of substance. They're also overly focused on beating competitors instead of creating value for stakeholders. Finally, perhaps the most severe attribute of a bad strategy is a lack of action. Instead, there are unreachable goals and lots of communication that don't lead anywhere. Rumelt summarizes the difference between a good and bad strategy in the way they help the organization achieve its goals. While good strategies have all the elements in place for achieving the goals, bad strategies don't. So, when defining a strategy, you know that the most important thing is to be able to achieve the desired goals.

Defining an API product strategy starts with an analysis of the current situation and an understanding of where you want to be at. Those are the two elements that help you understand what you need to achieve your desired goals. To paint the current scenario where you're at, you can do a market analysis. You start by doing a market segmentation where you identify groups of potential customers that might have similar needs. Those groups can be based on attributes that are relevant to your API product, such as their company size, the tools and services they already use, and their job roles. Those attributes help you identify smaller groups of people that you can analyze to understand their needs and how they'd use your API product. Another part of market analysis consists of studying what your industry is doing. As you saw in *Chapter 2*, your opportunities and challenges depend on the industry in which you're operating. Some industries, such as healthcare and banking, are heavily regulated, so that's something that you also need to pay attention to. Overall, a good market analysis helps you to understand the potential of your API product and how you can frame your unique value proposition to potential customers. To craft a good value proposition, you first need to understand the needs of potential customers and then find the key attributes that differentiate your offering from your competitors. At this point, and after you have a good sense of which market segments you'll want to target, you can identify your API user personas. They'll act as the voice of the customer and will guide all your API product design decisions going forward. As you've read before, measuring the progress of your strategy is fundamental to knowing if you're heading in the right direction. To do that, you rely on a set of metrics that reflect your strategy's goals. For business-related goals, you use metrics such as **monthly recurring revenue (MRR)** and **average revenue per user (ARPU)**. If your goals are more related to usability, you use metrics such as **time to first transaction (TFT)** and **monthly active users (MAU)**. Those metrics act as direct feedback from your API product to show you if your actions are leading you closer to your goals. If you still need to, you adapt your strategy, re-evaluate your goals, and redefine your metrics. Using metrics is just one way to get feedback. By engaging with stakeholders, you'll be able to understand how your API product strategy is performing. Keep reading to learn how to do that.

Stakeholders

By definition, stakeholders are anyone that has an interest or concern in the development and management of your API product. These can range from existing customers who have direct involvement in the API product to potential competitors who are observing your strategy and seeing how it can influence theirs. Your job is to identify, categorize, and understand each and every stakeholder.

Let's start by understanding how existing customers can influence the success of your API product. Customers are the primary users and source of revenue for your API product. The more customers you have, the more revenue you can potentially generate and the more you can learn to improve your API. Customers give you direct feedback about the quality of your API that you can use to adapt your strategy to reach your goals. If customers are on the user side of your API product, company shareholders or partners are on the investor side. Their needs relate to business factors and how your strategy is behaving to generate more revenue and profit that affects their bottom line. They will help you guide your teams to execute your strategy flawlessly while spending as little as possible. Speaking of teams, employees are also stakeholders. They have needs and interests as valid as other stakeholders. Employees contribute to developing your API product in the form of labor and receive a salary as a reward. They want to make sure that the API product is successful so that they continue to have a profitable relationship with your company. Other less-mentioned stakeholders are your suppliers. In your case, they are companies providing technology-related services and products you use to build and run your API. Suppliers are interested in the success of your strategy so that they can keep doing business with you. Among external stakeholders, you can also consider regulators and government agencies. These are less involved stakeholders; however, they sometimes can dictate your success. If you're operating in a heavily regulated sector, you should keep a close relationship with these stakeholders. Competitors are also a group of interesting stakeholders, especially when they're involved in the development of industry standards that you rely on. In the case of APIs, it's good to watch companies that contribute to OpenAPI, AsyncAPI, GraphQL, gRPC, and other industry specifications. They're interested in having the power to set the direction of API specifications, and that can change the success of your strategy. If possible, it's good to engage in collaboration with those stakeholders, even if they're your competitors. Overall, stakeholders can have a major influence on the success of your API product. As you've learned, different types of stakeholders interact with you in different ways. Customers use your API product and contribute to your revenue. Shareholders invest in your strategy with their funds. Employees invest in the future of your API product with their time and skills. Suppliers help you deliver your API by creating products and services that you use. Regulators create rules under which your API should operate. And competitors fight for a piece of your market share by setting the direction of industry standards. You can't control all stakeholders, but you should at least understand the needs and expectations of the ones that directly use your API.

Understanding what potential users and existing customers want is fundamental to building a winning strategy. That's because the best strategies are those that are aligned with what customers want. One of the ways of getting insights into what users are interested in is to ask them. Yes—asking users for their feedback has surprisingly good results. Most people are okay with spending some of their time sharing how they feel about your API product. In fact, many people enjoy the feeling of being part of the building process. It gives them a sense of ownership they'd only get if they were your employees. Several methods help you gather feedback. Let's start with methods that are proactive and then look at those that happen in reaction to something that happened.

The first technique is the most straightforward and involves nothing more than yourself, a stakeholder, and the will to have a conversation. I'm talking about user interviews. An interview is a one-on-one conversation with a stakeholder with the goal of understanding how they feel about your strategy. You can conduct interviews in person, over the phone, or online. However, interviews should be synchronous. Sending a list of written questions by email is not a good idea because you want to capture genuine answers that happen at that moment. Sometimes you don't have enough time to interview a large number of people, and you have to resort to other feedback methods. The next technique on the list is called a focus group. It's a way to get a small number of stakeholders together to discuss a topic related to your product. You pick a topic you want to learn more about and bring it to the focus group to get their perspective.

Similar to interviews, focus groups can be conducted in person and also online via video calls. Continuing with other feedback-gathering techniques, you can also do surveys. While you won't be able to obtain direct feedback or learn what stakeholders feel during an intense debate, you'll be able to quickly synthesize information. Surveys give you the power to reach a large number of people and analyze their thoughts in both quantitative and qualitative ways. Surveys are also easy to distribute and participate in. You can easily invite users to participate in a survey by emailing them. The goal is to make the participation fully asynchronous without any human intervention from your side so that it can scale to a large number of users. These are all proactive methods because they let you obtain feedback by explicitly asking stakeholders before they actually feel the need to share their feelings. However, you could also follow a reactive approach. That happens whenever you intercept communication from users regarding your API product to extract insights. One way to do that is to actively listen to what users share on social media. You can then store that information and translate it into meaningful feedback. Asking for feedback on every customer support interaction is another way to obtain feedback reactively. You can even measure the difference in the sentiment of users before and after interacting with the support team. Whichever method you follow, make sure that you can translate the feedback into actionable insights that help you make decisions.

The decision-making process is also something where you can directly involve stakeholders. If using feedback to make informed decisions is important, think about engaging stakeholders in helping you to make important decisions. While this might seem counterintuitive, empowering stakeholders can yield good results in terms of alignment of your strategy.

Let's explore two types of exercises that help engage stakeholders and plug them into the building of your API product. The first one is called consensus-building exercises. These types of exercises can be done in person or via a video call, as long as all users can freely participate. Some types of consensus building involve dot voting, affinity mapping, importance and difficulty matrices, and role-playing. According to the LUMA Institute, these techniques are a part of **human-centered design** (HCD). Dot voting is a way to understand, in a democratic fashion, the preferences of the stakeholders. You start by presenting all the options on a physical or virtual whiteboard. Then, you give each participant one sticky dot—or a square sticky note if you don't have dots. All participants vote on their preferred option simultaneously. In the end, you'll have a list of options ranked by order of preference. Affinity mapping lets you identify and group similar options to create clusters. You ask stakeholders to make sticky notes describing items related to the topic that you're studying. Each new sticky note is placed in proximity to other similar items. In the end, you'll have groups of notes that represent clusters of affinity. An importance and difficulty matrix is a way to rank items in terms of their importance and difficulty visually. This exercise helps you to rank your options when deciding what to build next. You start by drawing a two-by-two matrix with importance on one axis and difficulty on the other. Then, you ask participants to present one item at a time in the form of a sticky note. After that, you involve the group in discussing where to place the item. In the end, you'll see which items you should focus on next. Finally, role-playing exercises help you and your team empathize with your stakeholders.

You perform a role-playing exercise by first identifying what it is that you want to analyze. Then, you create a script that includes the tasks that you'll replicate. You'll perform those tasks as if you were the individual or group that you're studying. All the notes you take during the process help you better understand your stakeholders and, in turn, inform your strategy.

Another way of informing your strategy is to look at your business objectives. Let's take a deep dive into how your API product can support your business and how it can also generate costs.

Business objectives

Any successful product has to be aligned with business objectives. Otherwise, what you have is a set of features that don't generate any outcome favorable to your business. Understanding the business objectives that you want your API to support helps you during the API design stage. There are choices that you'll make that depend on how you want your API product to align with your business goals. Let's look at some possible business objectives to consider while designing an API product. The first, and perhaps the most obvious objective is to increase your revenue. When you think of a product, you immediately think of how you can make money with it. To generate revenue, your API product needs to add value to the customers using it. Without value, users won't want to pay for your API, and you won't be able to generate any revenue. Adding value can be done by providing features that solve problems that users have but also by using other tactics. Looking at what competitors are doing and making something better is one option to create value. That, in tandem with better product quality, makes users prefer your API over the competition. In other words, the value of your API product feels higher to your customers. Another option is to dramatically improve the user experience. Go back to what you learned in *Chapter 2* and apply it throughout your API product. Remember that your API can be used by a variety of users, not only developers, and pay attention to decreasing API friction as much as you can. Another type of business objective is not related to revenue but instead to costs. I'm talking about reducing some of your company's operational costs. An API product can help you reduce some of your operational costs in two ways. Firstly, the API can integrate with other services, and by using it, customers can onboard your product easier without any human intervention. Not having human intervention is related to the second way. Because APIs can be used programmatically, the operations they replace can be scaled in a cost-efficient manner. Other strategic goals might apply to your scenario. Market expansion is one of those goals where the API product can help your company reach new markets or expand its footprint on existing ones. Another possible goal is increasing customer retention. The more users integrate your product with tools they already use, the longer they'll stay with you. By offering an API that's easy to integrate with other products, you're indirectly increasing user retention. There are other business objectives that your API product can support, and your job is to identify them and make them a part of your strategy. Something else that you need to identify is the resources that your API product will consume.

Helping you reach your business objectives is a positive aspect of an API product. However, to do that, the API needs to be built and maintained. Along the API life cycle, you'll find that several types of resources will be consumed. Let's go through those resources to understand the impact of their consumption by the API product. Consumed resources can be split into people, infrastructure, and data. Starting with people, the impact happens mainly in the design and implementation stages of the API life cycle. However, pay attention to the sometimes hidden costs of maintaining and supporting your API. Initially, those costs are low or non-existent, but over time they will rise. Maintenance costs will have peaks whenever there's a new release or a fix to the API code. Support costs will keep growing over time until they become almost 100% of the total cost of the API product, as depicted in the following diagram:

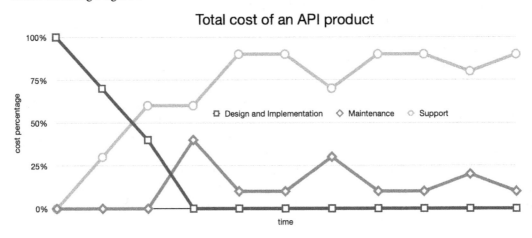

Figure 6.1 – Example of how the total cost of an API product evolves over time

Infrastructure costs are associated with the success of your API. In other words, the more your API product is used, the more resources it needs to scale. You start with minimum infrastructure, and you keep growing it as the usage of your API keeps growing. In that sense, if your API product monetization is related to its usage, you don't have to worry about infrastructure costs, as they'll be covered by the growth in revenue. However, if you're following the freemium model, watch out for how you're sponsoring all the free usage by paying for infrastructure. The other type of resource that your API product consumes is data. Your API needs to access and manipulate information located in different places inside your organization. While at first accessing data doesn't look like a cost, it might translate into an increased toll. For instance, if your API consumes an internal database, you want to make sure that the database can scale to the growing demands of the API or that all the logs and analytics that your API generates are stored somewhere. All that information consumes resources and has an underlying cost to your business. Knowing how your API product can influence your business objectives is essential. Another crucial area that dictates how you design your API is understanding the personas that will be using it. Let's now explore the relationship between personas and your API product strategy.

Personas

In *Chapter 2*, you had an introduction to user personas. You learned about lightweight or proto-personas, which are created based on assumptions. Now, you'll see how you can expand those initial assumptions into a set of qualitative personas. These are based on interviews and other research done on a small sample of your potential user base. Let's start by identifying the attributes of user personas. The first thing to always keep in mind is that personas are fictional characters. They're not meant to represent any single person or to identify a particular individual. The goal is to amalgamate traits and insights obtained from interviews with real people into a fictional character that represents the whole group without specifying anyone in particular. In other words, personas are generalized representations of a group of potential users or customers. However, being a generalized representation doesn't mean that you can't go deep into their attributes. The goal is to understand the whole group as much as you can and describe it by using a persona. Every time you interview someone, you add more information to the corresponding persona and enrich its informational value. That is why user interviews are so important to define an API product strategy. They provide contextual information that would otherwise not be available to you. While interviews are the preferred method for enriching user personas, sometimes they're not an option. In those cases, I recommend other types of direct user research, such as focus groups—following what you learned earlier in this chapter—and usability testing. Focus groups are better for understanding users' opinions about your API product. The exercise provides qualitative results not so much about the user group but more about how they feel about the topic being discussed. Usability testing, on the other hand, offers you the possibility of following what users are doing with your API. That lets you infer some of their attributes from their actions. You can understand what job someone is trying to complete by watching them work. Or, you can feel their challenges when they're trying to use your API product. Finally, you can clearly see the tools that they use just by observing them. Let's look at each of these attributes individually to see how they can map to your overall API product strategy.

Jobs-To-Be-Done (JTBDs) is a persona attribute that represents the jobs or tasks that the members of the group generally need to complete. JTBDs don't necessarily represent the job title of the user persona. Instead, they show the tasks that a group of people is generally responsible for. For example, a software developer persona might have a JTBD titled "integrate a payment gateway." The JTBD, in this case, reflects a type of task that members of the group of software developers that you're studying have in common. Identifying JTBDs can be as simple as asking people during interviews or more complex if you're observing how a group of people behave. Knowing what JTBDs are and how to identify them is the first step to understanding how they can influence your strategy. The tasks that a user persona needs to accomplish determine how your API will be used. JTBDs are a good way of understanding what your API will be used for, and that can determine how you define the different elements of your strategy. For instance, if you understand that most user personas will probably use your API only once to complete a one-off task, then offering a subscription payment model might not be the best choice for generating steady revenue. If, on the other hand, most JTBDs happen only once a week, then a pay-as-you-go monetization model will probably generate low revenue. Additional information can be determined from JTBDs, such as the importance of the task being completed and how your API can be critical to its completion. More critical tasks deserve more importance and have more value. If your API product is a key piece in completing critical tasks, it might be easier to sell to those user personas. In the end, understanding JTBD can open a range of insights that help you frame your offering to meet the needs of the personas you're targeting. Another persona attribute that can influence the shape of your offering is the set of challenges that personas face regularly.

In the context of user-persona analysis, challenges are anything that prevents cohort members from successfully completing their JTBDs. During your analysis, you might find different types of challenges, such as technical, time-related, resource-related, and regulatory. Let's look at each of those categories in detail and then see how they can help you decide which capabilities you'll offer in your API product. The first category, technical challenges, is related to the technical knowledge and skills that a persona needs to have to be able to complete their tasks. Each friction point you identify in how a persona handles their technical tasks is an opportunity for you to fill it with an API capability. Every time you see someone showing you how complicated a particular technology is, you can think of a capability that addresses that difficulty and turns it into something very simple to use. In the second category, time-related challenges, you'll witness complications related to how long tasks take to complete. Here, possible API product capabilities can introduce speed to task completion and address deadline constraints. Finally, the last category, regulatory challenges, is where existing government regulations make tasks hard to complete. You'll win the love of users if you can complete those same tasks while eliminating the burden of dealing with regulations. Doing it takes work. However, there's a high-value perception associated with it, and it can be a massive driver of revenue growth for your API product. The categories you've seen now are just examples of the types of challenges that you'll find when conducting research through user interviews. Whatever challenges you find, try to translate them into API capabilities. With time, you'll see that you can even group together challenges that can be addressed by a single capability. Another driver for enhancing your API product strategy is knowing about the tools that the personas use. Let's look at how that works and what information you can draw from that knowledge.

Tools are the services, applications, devices, and anything else sitting between users and the tasks they have to complete. In the context of an API product, knowing which tools personas use is very relevant because it gives you information about your API's compatibility. To be able to let users consume your API from within the tools of their choice, you need to make it compatible with those tools. Before that, let's understand how to identify tools during user interviews and what information is interesting to extract. While interviewing users, you can ask them how they normally go about completing their daily tasks. This is an open question that usually gives room for people to explain what are the tools they use to get their jobs done. You then take note of all the tools they mention so that you can subsequently ask for more information about each one. Two types of information are particularly important. One is how often they use the tool. Knowing how regularly someone uses a tool gives you a hint of the role of the tool in someone's workflow. The second type of information is related to the importance of the tool in someone's job. You want to know what would happen if the tool disappeared. This will give you information on how crucial the tool is to complete someone's tasks. If the same task can be done using other tools or by hand, then the tool isn't critical. If, on the other hand, there's no replacement, then it means that the tool must be used during a persona workflow. After knowing which tools are a must-have for each one of the personas, you can order them by importance. At the top, you place the tools that are used more often, and at the bottom, the ones with less use. Having this information is critical for the next step, which is deciding which architectural style to use based on the tools used by the personas.

During my work building different API products, I have identified a framework for deciding which architectural style I'd use based on the tools used by the different personas. As you've seen in previous chapters, API architectural styles are the technologies and protocols that dictate how an API operates and how it communicates with its consumers. Examples of architectural styles include **Representational State Transfer** (**REST**) and GraphQL, but there are more. The framework I use for identifying the right architectural style involves a prioritized list of tools that I described earlier in this chapter and a list of all the architectural styles that are feasible to implement. To begin with, you'll investigate each one of the tools to understand how they're compatible with the different architectural styles. You'll want to understand if an API using a certain architectural style can be consumed from inside each one of the tools. The consumption can be done directly or through some integration. The less direct a consumption is, the more friction it will introduce and the worse the user experience will be. As an example, if one of the tools is Google Sheets, you'll find out that it's possible to read data from an API using the REST architectural style with minimum customization. If, on the other hand, you're trying to create data from Google Sheets, you'll find that it's not possible to do it using the REST architectural style. After having the information for all the tools and architectural styles, you can draw a matrix and see the compatibility between styles and tools, as illustrated here:

	REST (GET)	REST (POST)	GraphQL	gRPC	SOAP
MS Excel	✓	VBA	Add-on		✓
Google Sheets	✓	Add-on	Add-on		
Postman	✓	✓	✓	✓	
JavaScript	✓	✓	✓	✓	✓
Web browser	✓	HTML forms	GraphiQL		

Figure 6.2 – Example of a compatibility matrix between architectural styles and tools

Not all tools may be compatible with one of the architectural styles, so you'll have to make a compromise. It's good that your list of tools is ordered by the most used. Deciding which architectural style to use is then an exercise of picking the one with the most used compatible tools and trying to find an alternative solution for the tools that aren't directly compatible. One thing is for sure—your choice of architectural style is based on what users need, and that alignment alone will make your API product user experience much better. Another area that helps you improve the user experience is understanding the behaviors of each persona while interacting with your API product.

Behaviors

After understanding the tools each persona uses and the capabilities you'll want to offer, it's time to learn about behaviors. Behaviors are what users do when they interact with a product. In your case, since you don't have a full API product ready yet, you need to understand how users interact with their tools of choice and then extrapolate the information to how they'd use your API. To do that, you'll need to ask users to show you, or explain, the steps they make using their tools to complete their tasks. Then, you overlap those steps with the capabilities you want to offer to get a list of the relevant steps. Those steps constitute the behaviors that you want to study deeper. You'll see that different capabilities translate to a different number of steps. Your goal is to analyze each one of those steps and learn about its value in the overall task that the user is trying to complete. You want to know if a step can be easily replaced or if a particular task has alternative steps based on certain conditions. Having as much information as possible about all the steps gives you the opportunity to translate them into features. Some of them will have a direct translation, while others will have to be grouped together to form a feature. Suppose you're studying the capability of making online payments without having to remember a credit card number. Among the steps, you would see that users would have to copy and paste a credit card number. That alone could translate into a feature. In this example, the feature would allow users to retrieve a credit card number from an existing list of stored cards. If user behaviors are crucial to identifying features, knowing how to identify such behaviors is critical.

One way to identify behaviors is to observe how users complete their tasks. You can have users with you in person and see what they do, or you can ask users to share their screens while performing workflows leading to finishing their tasks. This technique is often called fly-on-the-wall observation, and its key characteristic is that the observer—you, in this case—doesn't interfere with the user being observed. Using the fly-on-the-wall method, you can gather detailed information about a user's behaviors, interactions, and routines while interacting with their tools of choice. Following this observation, you can expand your findings by zooming in on particular areas of the behavior or zooming out to understand the whole picture. You can use a technique called abstraction laddering to go through different levels of abstraction and repeatedly learn why and how each behavior exists and how an API feature could potentially replace it. If users are unwilling to show you how they work, you can ask them during an interview. The value is not the same as with the fly-on-the-wall technique, but at least you'll get a description of each user's behaviors. The important things to identify and document are the interactions between different steps of each behavior. In particular, you want to know how users move information between the different tools and processes they use in their workflows. That will help you design API features that not only fit within the workflows that users are accustomed to but also work with their tools of choice. Deciding which features to build is not the only thing that you can learn from studying user behaviors. Keep reading to see what else behaviors can uncover.

Knowing that the features you're designing are what users want is excellent. However, not everything goes flawlessly all the time. Sometimes, users will need help completing their tasks using your API product. In those moments, you need to have a good support and troubleshooting system in place. So, knowing which support areas to invest the most in is fundamental. By analyzing user behaviors and workflows, you can identify which steps are more likely to cause friction and complications. It's expected that in many workflows, there are steps that are easy to complete, while others are complex and might require external assistance. During your behavior analysis, you want to focus on those steps where you see users struggling. Those are the ones that might make users contact your customer support or look for external help. That's how you decide where to prioritize your support and troubleshooting system. You focus first on helping users succeed in their most complex workflow steps.

Summary

At this point, you know what the strategy stage of the API design process looks like. You know how to identify your API stakeholders, understand the business objectives of the API, know about the API user personas and their JTBDs, and can describe their behaviors when interacting with the API. You now understand how defining a solid strategy will impact the success of your API product.

You started by learning the definition of strategy. Through a series of examples, you learned that identifying and achieving a specific goal is central to a strategy. You learned that having a clear strategy is essential because it provides direction, helps you to allocate resources, makes your decision processes easier, promotes accountability, and enables adaptation. Then, you got to know what the components of a strategy are, the goals you set for your API product, the specific objectives related to those goals, the tactics you employ to reach those objectives, the resources you need to follow those tactics, and the timeline where the action will happen. You learned the difference between a good and a bad strategy. In particular, you learned that a good strategy starts with clearly assessing the current situation. In contrast, a bad strategy has vague goals and is overly focused on beating competitors. After that, you learned about the different elements of defining an API strategy. You saw how to perform a market analysis, create a unique value proposition, identify your target market, define metrics, and create a feedback loop. Then, you learned about the different types of stakeholders, their needs and expectations, and how to engage them and manage the relationship. After that, you got to know how to define your API product business objectives and the costs it generates. You learned that there are different business objectives, not all related to increasing revenue. Then, you learned how to define user personas using techniques such as observation and interviews. You learned to identify the personas' JTBDs, challenges, tools, and behaviors. Finally, you learned how to decide which architectural style to use based on the tools used by the personas and which features to offer based on the behaviors of the personas.

Here are some of the things that you learned in the chapter:

- The definition of strategy
- How having a goal is at the center of a strategy
- A strategy provides direction, helps allocate resources, makes decision processes easier, promotes accountability, and enables adaptation
- The components of a strategy

- The difference between a good and a bad strategy

- How to perform a market analysis, create a unique value proposition, identify your target market, define metrics for your strategy, and create and maintain a feedback loop

- The different types of stakeholders and their needs and expectations

- How to define your API product business objectives and generated costs

- How to define user personas and their attributes

- How to decide which architectural style to use

- How to identify which API features to offer

Right now, you know that designing your API product starts with having a good strategy. To do that, you learned several techniques to identify, define, and document your API product strategy. In the next chapter, you'll apply what you've been learning by seeing what it takes to define and validate an API design. Continue reading to learn more.

7

Defining and Validating an API Design

Defining an API involves working together with stakeholders to reach a point where everyone agrees that the API will work well in a real-life scenario. This chapter will teach you the different techniques you can use to validate an API design. You will learn how to define the capabilities of your API by distilling the information you obtained during the strategy step. You will also learn how to use API mocks to prototype your API so that stakeholders can test it and reach a point where your definition is validated. You will learn that API mocks are fake versions or imitations of APIs. Mocks let you have an API that's ready to be tested without you having to build it first.

This chapter will begin by showing you what API capabilities are and how you can define them. You'll learn how to perform a use case analysis to define the functional requirements of your API product. You'll also see how business requirements, integration needs, security, legal regulations, and documentation play an important role in defining capabilities. After you've learned how to define capabilities, you'll learn how to create API mocks and why they're a critical piece when it comes to validating your API design. You'll learn about the different types of API mocks and, especially, how state management can be useful in situations where testing a chain of requests is needed. After, you'll see examples of how you can prototype an API integration with a UI to provide stakeholders an approximation of what a user-facing application would feel like. Finally, you'll see how iterating through consecutive versions of your API design is a part of the validation process.

By reading this chapter, you'll be able to analyze use cases to extract functional requirements. You'll understand how business requirements such as monetization influence the design of your API capabilities. You'll also know how to create and run API mocks to obtain feedback from stakeholders. Finally, you'll know how an API prototype UI can help you engage better with stakeholders and put you closer to validating your API design.

In this chapter, you'll learn about the following topics:

- API capabilities
- API mocks
- Prototyping an API integration with a UI
- Design iterations

Technical requirements

In this chapter, you'll come across different data types and formats. You'll also see some code examples of using JSON and JavaScript. Even though you don't need extensive knowledge of these technologies, it will help if you understand the basics of programming flow and have looked at a few JSON documents.

API capabilities

The visible part of the design of an API product is what users interact with. API capabilities are what users simply call "the API." Without capabilities, you wouldn't have an API to offer in the first place. That's why identifying and validating an API design starts by defining and testing capabilities. As you've seen earlier in this book, capabilities are derived from potential users' challenges. Let's learn how to structure the work of identifying capabilities by understanding use cases.

Use case analysis

Since not all potential users are alike, earlier in this book, you learned how to identify, define, and document user personas. If each user persona is an abstract representation of a group of people, then you can compile the things that the group of people does regularly. Among those things are the **jobs to be done** (**JTBDs**). I find the information about each JTBD particularly interesting because it lets you understand the goals of each persona. By identifying those goals, you can understand the benefits that your API product should offer. Those benefits, along with the identified challenges, can be directly translated into capabilities and then into features. Once you have these JTBDs and the goals and challenges of each user persona, you can work on defining use cases. A use case is a richer representation of a JTBD because it also describes the journey that users take to complete their jobs. Each use case has several attributes that help you identify it:

- **Title**: A name that describes what the use case is about and helps you quickly identify it.
- **Actors**: Any users, tools, or external elements that are part of the use case.
- **Goal**: The goal that the user is trying to achieve by performing the use case.
- **Pre-conditions**: Anything that needs to exist before the user starts performing the use case.

- **Flow**: A sequence of steps that users take to complete the use case successfully. You must detail each step as much as you can.

- **Post-conditions**: The expected outcome of completing the use case. These can be changes in information, systems, or even users.

- **Frequency**: A description of how often the use case occurs. Using a numeric score here will help you during use case prioritization, as you'll see in a moment.

- **Criticality**: A score of how critical the use case is for the user persona. You can follow a percentage score where 100% means that the user persona can't leave without the use case. The lower the score, the more redundant the use case is to the user persona.

To illustrate what a use case analysis looks like, let's look at an example while following the web payments capability that we've been referring to throughout this book. The following are the attributes of this example use case:

- **Title**: Online purchase with web payment

- **Actors**:

  - **Customer**: The person making the online purchase

  - **eCommerce website**: The platform where the customer selects and purchases products

  - **Payment gateway**: The service that securely processes the payment transaction

  - **Merchant**: The seller or business receiving the payment

- **Goal**: To securely and conveniently make online purchases by using a web payment system

- **Pre-conditions**:

  - The customer has selected the desired products on the eCommerce website

  - The customer has provided valid payment information, such as credit card details

- **Flow**:

  - The customer initiates the checkout process on the eCommerce website.

  - The eCommerce website presents the customer with payment options, including web payment.

  - The customer selects the web payment option.

  - The eCommerce website securely sends the payment details to the payment gateway.

  - The payment gateway verifies the payment information and ensures the transaction's security.

  - The payment gateway communicates with the customer's bank or payment provider to authorize the payment.

- The bank or payment provider validates the payment and sends the approval or denial status back to the payment gateway.

- The payment gateway informs the eCommerce website about the payment's status.

- The eCommerce website displays the order confirmation to the customer and provides any relevant information about the purchase.

- **Post-conditions:**

  - The payment transaction is completed successfully, and the customer's payment is authorized

  - The eCommerce website notifies the customer about the order confirmation

  - The merchant receives the payment for the purchased products or services

- **Frequency**: Multiple times per day, depending on the volume of transactions on the eCommerce website

- **Criticality**: This is very critical as it involves sensitive payment information, as well as the completion of online purchases

After identifying all the attributes for each use case, you can now prioritize them. Your goal is to eliminate any duplicates and come up with a short list of use cases that you can validate and potentially translate into capabilities. Use case prioritization is done by ordering the list of cases by how frequently they occur and how critical they are. The next step in defining API capabilities is getting feedback from users and iterating until you have a short list of use cases completely defined and validated. Preferably, you also want to be able to replicate each use case yourself. After that, you need to map each use case into product capabilities. To do that, you must translate the use case into a full capability and the steps of its flow into API features. You'll end up with a list of features for each use case that can be further refined and specified. One of the refinements that you'll do is how easy it is for you or your development team to implement each feature.

Feasibility is the attribute that captures how achievable it is to implement the feature. It is not just related to cost or time – it also has to do with other factors, such as technical difficulty, regulatory complexity, and operational frictions. You can identify the feasibility of each feature using a numeric score so that you can easily order the list of use cases. Your final output will be a short list of use cases, each with a list of features sorted according to frequency, criticality, and feasibility. The next step is to identify how users will interact with the features using your API product.

Functional requirements

Knowing how users will interact with your API is as important as understanding what you will implement and what you will choose not to build. Scope is a product attribute that you can use to communicate the boundaries of each feature in terms of functionality. For users, it's important to understand what each feature offers so that their expectations are aligned. For you and your development team, it's important to know what is being left on the table so that everyone is aligned and understands what needs to be built.

One thing to pay attention to is what is often called "scope creep." After you've identified and validated the scope of each feature, it's important to make sure you don't increase it. Otherwise, your actions can lead to implementation delays, increased costs, and misaligned expectations from stakeholders. Something else that can affect how stakeholders perceive your API is the user experience you're offering. In *Chapter 2*, you learned about API user experience and the hierarchy of needs. Now is the time to put that into practice. As you may recall, usability is at the bottom of the hierarchy, and it's the most important thing to care about to keep users happy. Offering good usability means that users can quickly and easily start interacting with your API with little or no friction. Right after usability, there's functionality or aligning what users need and what you're offering. Hopefully, you're covering that part well now. Then, you have reliability, which includes things such as performance, scalability, and error handling.

While performance is the ability to deliver results reliably under different circumstances, scalability refers to being able to quickly adjust your infrastructure to meet changing user demand. Scalability ensures that your API's performance stays the same no matter how many users and what type of usage it has. Error handling enhances reliability in the sense that it provides a degree of comfort to users in situations where things don't work as they expect. Without proper error handling, users wouldn't feel that they could trust the API to perform the use cases they care so much about. If you go up the hierarchy of needs, you'll see other items that you need to care for. Your goal is to address each one of the hierarchy layers and make sure you're providing what users need. Another item worth mentioning related to functional requirements is the ability to test your API features consistently.

This is a good time to define and write down the functional tests that your API will have. These functional tests are particularly important because they guarantee that the functionalities that you define for your API product are running. They should verify that the functional elements of the API are all working, as well as validate user input and the type of response that users would get in a real-life scenario. Finally, functional tests should also verify situations that produce errors to mimic what users would see in those scenarios.

Business requirements is another area worth paying attention to when defining and validating your API design. If functional requirements were about understanding what users get when they interact with the API, business requirements have to do with business objectives that the API must meet to be viable. In many cases, the first business objective that you will think of is revenue. If the API product is to be commercialized, a monetization plan has to be put together. You have different monetization options at your disposal, as you learned in *Chapter 3*. Your goal now is to align your business requirements with a monetization model that helps you reach your objectives. Think about whether you should use the freemium model, the tiered model, or the pay-as-you-go model for your business requirements. You don't have to implement the monetization model right now. However, it would be best if you documented it in a way that makes it easy to put it into action when the time comes. Another area that needs to be documented is related to the analytics and metrics that have to be implemented to guide your business strategy. Here, you document each needed metric and explain how it helps the business understand how it's aligned with the overall strategy.

Metrics related to operational costs are particularly important from a business perspective because they help you understand the viability of the API product over time. Something else that helps the API align with the business over time is following any existing internal policies. If the API is operating in concert with the rest of the business, the chances of it succeeding are higher. To meet that requirement, you can use a tool called API governance. By governing the behavior of your API, you're making sure that it aligns with existing design standards and that its life cycle follows an approach already established in your company.

When you're creating a prototype, it's fine to do things your way and be as quick as possible. However, when you're productizing your API, it's more important to make it work seamlessly in the ecosystem that already exists in your business. By documenting your API governance approach, you're indicating how you intend to enforce the practices that the rest of the organization considers sound. Without a good governance program in place, things such as integrating your API with other systems will appear complicated.

Integration needs

Integration is yet another domain that you want to pay attention to. Your API needs to work with other internal – as well as external – systems and data sources. Understanding and documenting all those connections is important at this stage. You start by getting the definition of the capabilities you identified earlier and extracting the objectives of other systems. You want to know how each capability needs to communicate with other APIs, data sources, and any other system to provide its functionality to users. As an example, the capability of making payments that we've been referring to throughout this book needs to communicate with a database of credit cards and a payment gateway, among other things. The objectives of the payment capability are to obtain the user's credit card information from the database and then use it along with other payment information to make a request to the payment gateway. Identifying these objectives is fundamental to understanding and documenting the integrations that need to take place between your API and other systems.

In this example, your API would need to read information from the credit card database and also make secure HTTP requests to the payment gateway. The number of systems in your organization determines the possibilities of augmenting your API product by integrating it. To be able to take advantage of all the existing functionalities that different systems provide, you can document a landscape. The system landscape provides a high-level view of what is possible and acts as a catalog of existing features that you can leverage when you're implementing your API product. Something else that will surface from analyzing the system landscape is the existence of integration patterns. These are ways of integrating different systems that look similar to one another. After some studying, you will understand that many familiar patterns can easily be used by your API product. Interestingly, access to all those systems is usually restricted to prevent access from unauthorized parties.

Security and access control

Access control is one key area that you should pay attention to. Controlling who can access your API prevents unauthorized requests that can damage your business and generate unwanted loads on your internal systems. This is especially critical if your API lets users manipulate information or read business-critical data. You don't want private information to be openly accessed by anyone. You also don't want anyone to be able to change information that is stored in the databases that the API accesses. Even if you don't know whether your internal systems and data need protection, it's always better to be safe than sorry. At this point, you can document a simple access control system that can later be reviewed and amplified if needed. You should document different access control attributes, as follows:

- **Authentication**: How the API will authenticate its users. Popular authentication mechanisms include API keys, OAuth, and **JSON Web Token (JWT)**. Each authentication method has its advantages and disadvantages, all of which you should study to decide which one to use.

- **Authorization**: How the API will control what actions and resources users can access after they're authenticated. **Access control lists (ACLs)**, **role-based access control** (RBAC), and **attribute-based access control** (ABAC) are known methods. As a rule of thumb, using a method similar to what other systems in your organization use is a good idea.

- **User management**: How user information will be stored and managed and who will own the information inside your organization. To understand user management, you can document common functionalities related to user sign-up and other administrative tasks such as updating the authentication credentials.

- **Secure communication**: How secure is the way information flows to and from the API? Even though it's a common practice nowadays, it's a good idea to specify a secure communication approach by documenting it. HTTPS is the default secure communication for web-based APIs, and most engineering teams can easily implement it.

- **Rate limiting and throttling**: What happens when users try to make too many requests too quickly? To prevent abuse and overloading your API, as well as the internal systems it connects with, you should document your approach to limit user requests under heavy-load situations.

Compliance with laws and regulations

The final item on the list deserves more introspection. I'm talking about regulatory compliance and legal considerations. Depending on your industry, you're going to run into regulations. In most cases, governments impose ways for you to operate the information you obtain from users. Making sure that your API access control and security policies are aligned with existing regulations is mandatory. Otherwise, you can run into legal issues that will prevent customers from effectively using your API. One of the most well-known regulations is the **General Data Protection Regulation (GDPR)**. It's a comprehensive data protection and privacy regulation that was created by the European Union in 2018. The goal of the GDPR is to protect the privacy and personal data of citizens and make sure that all European Union states follow similar data protection laws. Even if you're not in the European Union, you should pay attention to GDPR as some of your customers might be affected by that jurisdiction. One of the key aspects of GDPR is that you should obtain consent from users before storing – or processing – their data. But there's much more to it than this. Because complying with GDPR is critical for the success of your API product, I have outlined some of the things that you should know about:

- **Scope**: GDPR applies to organizations that handle information about users residing in the European Union. This is regardless of where the company is located. If at least one user is in the European Union, GDPR is applied.

- **Consent**: Before processing users' information, you need to obtain their permission. Additionally, you must have a lawful reason for having to process their data. Acting on a user's behalf is considered a lawful reason.

- **Users' rights**: Users have special rights when it comes to accessing, amending, and even deleting their data. You must implement ways for users to execute their rights.

- **Cross-border data transfers**: Moving information from the European Union into countries without adequate data protection is restricted by GDPR.

Overall, GDPR has major importance in the design of your API product. Without good care, you can run into situations where you fail to comply with regulations and might run into legal complications. Another type of regulation is the **Health Insurance Portability and Accountability Act (HIPAA)**. Complying with HIPAA is only required if your API product manipulates health-related information. HIPAA has similarities to GDPR in the sense that one of its goals is to protect users' data and their ability to modify it. I won't cover it in detail here because its scope is limited to one industry.

Documentation

The final topic to consider in the field of API capabilities is documentation. You learned about API documentation earlier in this book, so let's cover the important aspects to consider when documenting capabilities. First of all, all the elements mentioned until now have to be documented. While some elements can be considered private, most of them must be available to all the stakeholders. For example, every stakeholder needs to understand how API authentication works, but only stakeholders internal to the company should know the details of how authorization validates a user's ability to perform actions. The important thing is that each API capability should be fully documented, first in the form of a reference and then in the shape of a tutorial that guides users. The goal of the tutorial is to be something that users can follow to learn how to use the API. The purpose of the reference is for it to be quickly searchable, and its content should be easily consumed.

Alongside the reference, you should include code snippets or even a fully formed API client in the programming languages that users are most familiar with. Offering an easy way to start using your API improves its usability, which is the first layer of the API hierarchy of needs. Another area of documentation that is often overlooked is related to changes. Having an always up-to-date changelog helps stakeholders understand how the API product evolves and make comparisons between different versions. Finally, adding examples to the API reference not only helps stakeholders better understand how each capability works but also helps you create and manage API mocks. Keep reading to see how having an API mock server helps you validate your API design and shortens the stakeholder feedback loop.

API mock

The goal of an API mock is to offer a simulation of an API so that stakeholders can test it before it's implemented. API mocks include simulated API requests and responses. This lets users interact with a simulated API to see how it behaves and how well it integrates with their existing tools. There are many benefits to offering an API mock, both for yourself as the API producer and also for API consumers. An API mock makes it easy to set up integration tests early on, even before the API is fully implemented. It also provides the opportunity to create prototypes of applications that use the API. It's an add-on to the documentation you provide, as it lets users see how the API works without them even having to sign up. Finally, it helps you whenever you want to release a new version of the API. By having a mock of the new version, you can test its compatibility with existing users and tools and check whether there are any breaking changes with the previous version. Overall, an API mock is a great way to obtain feedback from stakeholders when you're validating an API design. By testing the API through its mock, stakeholders can share their feedback, and you're able to iterate on the design progressively. Let's learn how to create an API mock and what the available techniques are.

Among the available API mocking techniques, my first recommendation is to do it manually, if possible. If your API output format is not binary, you can create mock responses yourself. Creating responses might be as simple as creating a text file using XML or JSON to format the response. Then, you can make that text file available on a web location that can be accessed by your stakeholders. Whenever they make a request to the URL where you stored the text file, they will receive its contents. It's as simple as that, and it's one way to have an API mock server with minimal setup.

The second technique that still doesn't involve any coding is using a tool such as Postman or Stoplight. These tools let you create a representation of the capabilities that you want to mock and generate a mock server that can be accessed by stakeholders. These tools work well with REST APIs.

However, other available tools on the market offer similar functionality for other types of APIs. The interesting thing about these kinds of tools is that they let you quickly put an API mock together and iterate quickly without having to write any code.

If writing code is not a problem for you or if you have a team of developers that can help you, then the next option is to create a very simple API server to be used as a mock. The API server can be built in a way that, while simple, can be evolved into the end solution after the API definition has been fully validated. There are many options, and I recommend using a framework. Popular frameworks for well-known programming languages include FastAPI for Python, Koa.js for Node.js, and Laravel for PHP.

Your choice of programming language and framework depends on what you or the developers you work with are familiar with, as well as the current practice in your organization. Whatever you choose, your goal is to be able to get started quickly and have a way to iterate easily so that you can incorporate any feedback that you receive into the mock until the API design is fully validated by the stakeholders.

As you can see, API mocking is mostly about creating responses that represent what you're going to implement. To create well-crafted responses, you must start by designing a model of the data for each response before creating the fake data that the mock server will use. You must start by thinking about the data structures that you want to use. Structuring the data means that you need to define the types of data and how it's organized. Depending on the response format you're using, you have access to common data types that can be used to represent your information. Let's look at some of the most used data types to understand what you can do with them:

- **String**: Used to represent a piece of text of any shape or size. You can use a string to define a name, an address, or even something more elaborate as a message.

- **Number**: This is used to represent any numeric value. You can use a number data type to store integers or floating-point values.

- **Boolean**: This is used to represent values that can only have true or false states. Boolean values are typically used to convey whether attributes are turned on or off. For example, you'd use a Boolean value to indicate whether a credit card is blocked.

- **Object**: Objects are used to define a collection of values, each one defined by a distinct key. They can be used to group attributes of the same element. As an example, an object representing a credit card would have stored its number, expiration date, the cardholder name, and any other attributes. Even though these attributes could be defined individually, having them together as a single entity is more practical and easier to use and understand.

- **Array**: These are used to define a list of items. The items can be similar or they can be of a different type. However, it's a good practice to ensure all the items are of the same type.

Now that you know about some different data types, you can use them to create responses that the mock server will use. Independently of the API architectural style and the format you choose for your responses, you'll have to make sure that stakeholders can use those responses as part of their JTBDs.

One key aspect to consider is evaluating the best response format to use. The API architectural style you choose will influence the data format, but there are other factors that you should consider. Let's start with the most popular data formats and how they map to different architectural styles. There are two main groups of data formats to consider: you have text-based formats, which are easy to read, even without the help of any tool, and you have binary or serialized formats, which aren't easy to read without a tool that understands them. Text-based formats should be used in situations where people must be able to read your content easily. They're also the ideal choice for interoperability and integration because many tools and programming languages can work with them. Because of their pervasiveness, text-based formats are widely used on the web. In particular, web browsers use HTML and other text-based data formats to transfer and manipulate information. Even with all these advantages, text-based formats aren't ideal in different situations. That's why binary formats are also a great choice. Binary formats excel in situations where efficiency and performance are critical. Binary formats generate data that's more compressed than its text-based alternative. Smaller data takes less time to transfer and interpret by the API consumer. Another factor regarding binary formats is that they can easily provide data confidentiality. Encrypting data is easier if you're using a binary format. Finally, because of its smaller footprint, binary formats are ideal in situations such as IoT, where the cost of transferring large amounts of information is prohibitive. Now, let's look at some of the most popular formats available for you to use, both text-based and binary:

- **JSON: JavaScript Object Notation (JSON)** is a lightweight text-based format that is human-readable, simple to understand, and widely supported by different programming languages and tools. It's used commonly by web APIs that follow the REST architectural style.

- **XML: eXtensible Markup Language (XML)** is a versatile text-based format that can be used to represent and structure data. Historically, it's one of the most commonly used text-based formats in APIs. It became popular with the SOAP architectural style.

- **CSV: Comma-Separated Values (CSV)** is a very simple text-based format that's widely used to represent tabular data. It's commonly used as a way to import and export data to and from a spreadsheet.

- **YAML: YAML Ain't Markup Language (YAML)** is a human-friendly and easy-to-understand text-based format. Additionally, it can be manipulated by most programming languages.

- **BSON: Binary JSON (BSON)** is a representation of JSON documents in a binary way. This format extends the capabilities of JSON by offering additional data types and efficient binary encoding. It's used by MongoDB, a popular database system, to communicate with clients.

- **Protocol Buffers**: Also known as **protobuf**, this is a binary format that can be used by most programming languages. Protobuf offers a compact binary serialization of structured information. This format is perfect for high-performant APIs.

- **Apache Avro**: This is a binary data serialization format that's widely used with asynchronous APIs. It offers support for rich data structures that can evolve.

This is just a taste of the data formats that you can choose for your API product. While setting up your API mock, it's also important to think about the contents of the responses that you're creating.

One way to make an API mock feel more like a real API is to provide responses that are not static. In other words, if stakeholders always receive the same response, they feel that they might be missing important aspects that should be tested. To provide different responses and cover a reasonable range of what the data would look like in a real-world scenario, you can create dynamic mocks. This type of mock is based on the same foundation as static ones. However, instead of delivering static responses, it generates a different response according to different factors:

- **Time-based**: The dynamic mock can generate a different response after a certain amount of time.

- **Request-based**: The dynamic mock can generate a different response for every request that's made.

- **Parameter-based**: The dynamic mock can generate a different response for every combination of parameters sent with the request. In the case of HTTP, headers are also included.

Different responses are often generated by the mock server by either rewriting the whole payload or by replacing certain parts of it with values that change according to one of the criteria mentioned previously. While the former approach is easier to understand, the latter offers the most versatility. Mock servers can identify certain parts of the response as variables and dynamically generate new values that are sent in the response payload. If you're using text-based responses, you can use template markers in the mock configuration so that it knows which parts it's supposed to replace and which ones it will leave static. If, on the other hand, your responses are binary, you can configure which parts of the response are dynamic by identifying them on the mock server configuration.

Another aspect to consider when making mock responses dynamic is what values the mock server will generate. You wouldn't want to have a date attribute dynamically mocked as a number and vice versa. Mock servers that support dynamic responses should offer the possibility of configuring the data type and the algorithm used to generate different responses. The algorithm can be as simple as randomly generating different responses to obtaining a different response based on different criteria, such as reading values from an existing list. Random responses make the most sense when what you're trying to validate is the shape of the response and how users would interact with it. It doesn't matter what information is delivered in the response as much as what the response shape is. However, if you want to offer mock responses that feel realistic, you need to offer more than random responses. Something else to take into account is how you can handle a sequence of requests to fulfill a use case. That's when state management enters the scene. Keep reading to learn more.

As you've seen, API mocking can start with very simple, static responses and grow to involve more elaborate dynamic values. The final stage of mocking is being able to handle chains of requests that are related. The information that's delivered in any response needs to make sense in the chain of requests. This can be done by managing the state between requests and persisting the data that's being sent by users. As an example, consider a list of requests that involves creating a payment and then checking the balance of an account:

1. **Create a payment**: This request receives an identifier of the created payment and subtracts the paid value from the user's account.

2. **Read account balance**: This request retrieves the last known balance from a user's account identified in the request. The account balance should reflect any recent changes.

The second request needs to be aware that a payment has been made in the first request. To create that awareness, mock servers have to be able to persist information in between calls to the server or, even better, offer a full mock environment that resembles what the whole system would do in production. While the goal is not to replicate a full production environment, you should be able to test sophisticated capabilities involving chains of requests.

The final topic of API mocking that you should pay attention to is collaboration. Whatever mock solution you use, make sure that it lets you collaborate easily with the stakeholders that are testing your API. Being able to receive timely and in-context feedback is critical when you're trying to validate your API design. Most of the API mock tools offer a feature that lets you at least add comments or share messages with other participants. If your tool of choice doesn't have this feature, you can always rely on email or the project management tool that you're using. The information you share with stakeholders has to be clear and concise to avoid misunderstandings. Your primary goal is to obtain feedback from stakeholders, so you must evolve the design of your API. You want to inform stakeholders how they can test the API using the mock and then ask them how they feel about it. The topics can help you understand what areas of your API need to be revisited:

- **User experience**: Understand how easy your API is to use and what improvements a stakeholder would add to it.

- **Capabilities**: Learn whether the API is meeting the needs of the stakeholders or whether there's anything that needs to be changed in the offered capabilities.

- **Integration**: Verify whether the API, in its current form, can easily be integrated with the tools that the stakeholder is using. If not, learn what changes you need to introduce.

- **Data formats**: Get feedback about the formats you've chosen for the API requests and responses. Verify whether stakeholders can consume – and understand – the format that the API is using.

- **API structure**: Learn whether the structure and naming used for the API is clear to stakeholders.

After you get feedback on all these questions, you'll understand whether you need to revisit your API design and what changes you have to introduce. If that's the case, then you need to ask stakeholders to test the improved API mock and provide more feedback. You must repeat this process until you reach a point where the feedback is positive and you feel you have an API design that you can use to build a prototype. You'll learn how to create a prototype in the next section.

Prototyping an API integration with a UI

At this point, your API design has been validated, and you're ready to implement it. Almost. Before that, I want to show you that you're still missing one type of validation that might reveal that the API design still needs to be refined. By creating a prototype of API integration with a UI, you're able to put it in front of non-technical users to understand how they interact with it. I'm going to show you different tools and techniques to create UI prototypes that can connect to your API mock and interact with it. Let's start with an easy approach that doesn't require any code and uses a tool that lets anyone build web applications. Retool is a popular web application creation tool that can connect to an external API that uses the JSON response data format. With this combination, you can put together a visual UI that loads data from your API mock. The second approach uses Postman and one of its features, called Visualizer. With Postman Visualizer, you can put together a visual representation of any API response. Whenever someone makes a request to your API mock using Postman, they'd get a visual response instead of seeing the response payload. This visual is totally up to you to choose and put together. It's based on HTML and can include dynamic JavaScript code that lets you create an interactive UI. While this approach is simple, it requires some code leveling. Finally, the third approach is to build a simple dynamic web page that consumes information from your API mock. While it requires coding abilities, it provides the biggest degree of freedom. Let's start by learning how to use the Retool approach.

Using Retool or some similar tool doesn't require any coding knowledge. However, you need to know web design concepts to be able to put together a usable prototype. For this example, you'll create a simple credit card payment form to simulate making a payment, as well as a credit card balance information screen to simulate displaying the account balance after a payment is made. Whenever someone submits a payment using the payment form, a request is made to the API mock's payment capability. Then, another request is automatically made to the API mock's balance capability to refresh the available balance. Let's learn how to put this all together. First, create a form with the following credit card information fields:

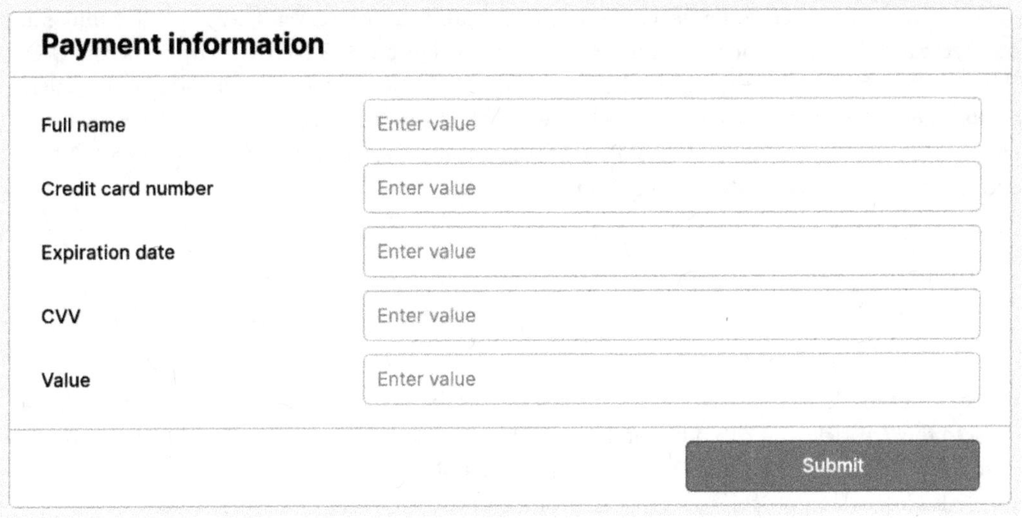

Payment information

Full name	Enter value
Credit card number	Enter value
Expiration date	Enter value
CVV	Enter value
Value	Enter value
	Submit

Figure 7.1 – An example payment information form created using Retool

Next, link the **Submit** button to a request to the payment endpoint on your API mock. Whenever someone clicks on the **Submit** button, a request will be made to your API mock, sending all the information from the form. Your API mock should react with a response indicating that the payment was accepted. Now, create a component where you'll display the information about the account balance:

Account balance: $1,234.56

Figure 7.2 – Example account balance information displayed using Retool

Add a trigger to the **Submit** button so that when someone clicks it, it will also request the balance information endpoint on your API mock. Use the information you received from the balance information endpoint to update the account balance component.

And there you have it! With this very simple prototype UI, stakeholders have a feeling of what it's like to use an application that consumes your API. Now, let's learn how to use Postman Visualizer to achieve a similar effect.

To use Postman Visualizer, you must create a collection and a mock server. Create a collection and, inside it, create a request named **Account balance**. This will be an HTTP GET request, and the URL will be `{{baseUrl}}/accounts/{{accountId}}`. Notice the values inside double curly brackets. These are variables that will be replaced with values stored in an environment. Now, create an example inside that request and name it **Balance for account 1**. Replace the URL for `{{baseUrl}}/accounts/1` and add a JSON response body:

```
{
  "id": 1,
  "balance": 1234.56
}
```

This request example will be used by the mock server as the response whenever someone makes a request to the `/accounts/1` endpoint. Now, create a mock server and use the collection you just created as its collection. Copy the URL of the mock server and store it in an environment variable named `baseUrl`. Now, whenever you select that environment, you can make requests to `{{baseUrl}}/accounts/1`, and the response will come from the mock server. Great! Now, let's add the visualizer to the request. To do that, open the **Tests** tab and add the code that creates the HTML template and sets the data for the visualizer. Let's also format the balance value to make it look like a currency:

```
const template = `
<h1>
  Account balance: <strong>{{balance}}</strong>
</h1>
`;

const response = pm.response.json();
const balance = response.balance.toLocaleString('en-US', {
    style: 'currency',
    currency: 'USD'
});

pm.visualizer.set(template, { balance });
```

Now, make another request to the `/accounts/1` endpoint and click on the **Visualize** tab. You'll see a result similar to what we obtained using Retool. The interesting thing now is that you can create more examples with different balance values and see the result of the visualizer adapt accordingly. At this point, you can add more examples for other requests and create more visualizers, or you can start building a complete HTML-based prototype. What matters is that you can share the API mock with stakeholders to obtain their feedback. Every piece of feedback you receive helps you iterate through the design stage of your API. Keep reading to learn how to manage these iterations and reach your goal of validating your design.

Design iterations

If you think your job is now done, you're wrong. What you've done is just one of possibly many iterations that you'll go through. The goal of iterating is to improve your API design by enhancing and polishing it with the feedback that you receive from stakeholders on each iteration. However, it's always good to be careful about the credibility and relevance of the feedback you receive. One thing to keep in mind is the level of expertise of the person giving you feedback. You can build a profile of each stakeholder where you identify their experience level, domain knowledge, and familiarity with API products similar to the one that you're building. These factors will help you understand which stakeholders are credible so that you can give more importance to their feedback.

Another way to check for feedback credibility is to verify how a stakeholder tested your API product and compare their actions to their feedback. If the provided feedback is aligned with their actions, that means that it's credible. Otherwise, there's a mismatch between words and actions that needs to be investigated. Even if the feedback you receive is credible, you should also check whether it's relevant. The best way to understand how relevant the feedback is is to compare it to what most stakeholders are expressing. The most relevant feedback relates to what most stakeholders are experiencing. Knowing how to identify credible and relevant feedback is as important as prioritizing it.

You can make every API design iteration count by understanding what to improve based on the feedback you've received. To do that, you need to be able to categorize and prioritize all the requests for improvements that you received during the iteration. Categorization is done by identifying common themes across all feedback and using those themes to create groups of items. Then, you can analyze each category and evaluate its feasibility. In other words, you must document the effort it would take

to address each of the categories. Effort becomes one of the vectors of prioritization. The second vector is the impact each category has on the strategy of your API product. There will be categories with high impact and low effort, others with almost no impact, and everything in between. If you assign numeric values to effort and impact, you can place each feedback category on a chart and start seeing which ones you should work on first. Those with a high impact are critical to address. However, from those, you can first work on the ones that have low effort. That is how you prioritize the feedback that you receive and make it actionable.

Now, you can go back to the beginning of this chapter and take the list of prioritized feedback as the items that you'll be refining. Some of the feedback will translate into polishing capabilities, while other feedback will turn into creating brand-new capabilities or even abandoning others. What matters is that you progress toward the goal of having an API product that aligns with the stakeholders and has a high chance of being successful.

Summary

By now, you're fully aware of what defining and validating an API design entails. You now know how capabilities are important ways of describing what your API product offers. You know how to do a use case analysis, define a user persona, define the scope and prioritize requirements, identify and document integration patterns, create and use API mocks and UI prototypes to obtain feedback, and iterate through your design until it's validated.

You started by learning about the relationship between API capabilities and an API definition. You learned that understanding use cases is fundamental to how you define your API capabilities. You got to know how to define user personas and their goals, objectives, and scenarios. You also learned how to describe and prioritize use cases and how to map the information into API product capabilities.

Then, you learned what functional requirements are. In particular, you learned about feature scope, API user experience, performance, scalability, error handling, security, regulatory compliance, and testability. You also learned that business requirements can determine the viability of your API product if functional requirements are essential. Then, you learned how to define API monetization and the analytics and metrics to measure its success over time. You also learned about API governance and how identifying and applying internal policies can help your API succeed in the existing organization landscape.

Next, you learned that understanding how your API can integrate with the tools that consumers use increases your chances of succeeding. You learned how to identify integration objectives, the system landscape, internal and external patterns, and potential data transformation.

After that, you learned about the different security and access control mechanisms that you can use. Following this, you understood how to identify and document the existing compliance and legal considerations to prevent your API product from being shut down. Then, you learned why having an API mock is a critical part of validating an API design. You learned about existing mocking techniques, how to model your request and response data, and how to choose a suitable response format to use. After that, you learned what dynamic mocking is and when to use it. You also learned about state management and data persistence to make an API mock more realistic and obtain better feedback from stakeholders. Then, you learned how to create a UI prototype using an API mock. You got to know that having a UI prototype increases the quality of the feedback you receive from stakeholders. Finally, you learned how to iterate through the API design. You learned how to identify user feedback credibility and relevance and how to prioritize it and translate it into API product improvements.

These are some of the things that you learned:

- The role of capabilities in the design of an API
- The relationship between use case analysis and API capabilities
- How to define a user persona
- The value of user goals, objectives, and scenarios
- How to describe and prioritize use cases
- The role of functional requirements
- How to define scope properly
- How business requirements affect the design of an API
- The internal and external integration needs of an API
- How to identify and document integration patterns
- The purpose and benefits of API mocks
- The existing mocking techniques
- How to model data to offer meaningful mock responses
- Dynamic mocking and state persistence
- How to use UI prototyping to gather end user feedback
- Iterating through the API design until it's fully validated

At this point, you know that an API design is only finished when the feedback you're receiving from stakeholders is positive. You learned about various techniques you can use to define and align your API product with the needs of users and the business. Additionally, you learned how to engage with stakeholders while providing them with resources to help them express their thoughts and give feedback on the API design. You're now ready to start specifying your API, which you'll learn about in the next chapter.

8
Specifying an API

Specifying an API is more than converting your design into a machine-readable format. This involves applying what you've been learning to create something developers can use to build and consume your API. You'll see that there are many API types and specification formats that you can use, depending on what your objectives are. You'll learn how to connect the discoveries you made while interviewing user personas with your business objectives and technical constraints, allowing you to decide which API architectural style to use. Then, you'll be able to actually create a machine-readable API definition document and apply API governance rules that fit your objectives.

This chapter starts by explaining how to choose what type of API to use from the different types showcased in *Chapter 1*. Identifying the correct type of API has to do with the behaviors and capabilities previously defined. After choosing which type of API to use, you'll see what each API specification format looks like, and you'll be able to compare the differences between formats. You'll be introduced to an example of a definition for the same API, written on each one of the specification formats we're covering. Finally, you'll understand how API governance can positively influence the quality of your API by promoting consistency of design and life cycle management.

After reading the chapter, you'll know how to choose an API type based on user needs, business objectives, and technical constraints. While doing that, you'll gain deep knowledge of the different types of APIs and what they're best suited for. You'll also see what the different API specification formats look like. After that, you'll have the chance to go through the machine-readable definition of the same API using several popular specification formats. With the definition documents in hand, you'll learn what API governance is and how to apply it to your design process.

These are the topics that you'll explore in this chapter:

- Choosing the type of API to build
- Creating a machine-readable definition
- Following API governance rules

Technical requirements

In this chapter, you'll be getting familiar with documents written in JSON, IDL, GraphQL schema, XML, and YAML. You should have a minimum understanding of operations, objects, and attributes and be able to identify their representation in a text file.

Choosing the type of API to build

Knowing what to build is the first step to success. Without knowing what type of API you're building, you'll end up choosing something that doesn't fit the needs of your users. Let's start by looking at API architectural styles, following what you've seen in previous chapters. There are multiple ways to pick the right architectural style for your API. In *Chapter 7*, I introduced you to a framework that I've been using that connects the tools that personas use to the architectural style of your API. However, you can also consider other factors in your choice of architectural style. We'll review your different options to identify the right architectural style to use, but first, let's recap the concept of architectural style.

In API terminology, you refer to its architectural style whenever you want to refer to the design principles, patterns, and constraints behind the structure and behavior of an API. You can use architectural styles to name commonly accepted architectures that APIs use. By using the same naming conventions, people understand each other and know exactly what they're talking about. Some examples of popular architectural styles include REST, gRPC, GraphQL, and, on the asynchronous side, **Event-Driven Architecture** (**EDA**). Each architectural style has its characteristics and trade-offs, so choosing the right one to follow is crucial.

Because identifying the right architectural style is so critical, you want to align your choice with three areas. Firstly, you should find alignment with the needs of users. If you don't address your users first, your API won't have any meaningful audience and will be challenging to consume. Not having engaged users leads to poor product retention and long-term failure. Secondly, you should align with the business objectives of your company or organization. Even if you have the best API that meets users' requests, it will fail if it doesn't help your business reach its objectives. Whatever the strategy behind your API is, now is the moment to ponder how the right architectural style can help it succeed. Finally, you should align with any technical constraints within your company or organization. Again, having an API that aligns with users and business objectives but goes against your company's technical practices will set you up for failure.

Aligning the API architectural style with the needs of potential users is done by following the framework I presented before. While identifying the user personas, you catalog the list of tools that they use regularly. Your goal is to ensure that your API can be easily consumed from within those tools. If users have to change their routine just to use your API, they won't find it easy and will probably give up after a few tries. By contrast, if they can use your API alongside the tools they're already using, they'll find it natural and simple, and they'll keep using it over time. You'll probably find more than one architectural style that can be a good match. Also, you'll realize that not all architectural styles can be used directly from all the identified tools. It's expected that you'll have to find add-ons or plugins just to make the connection between specific tools and your API work. Create a list of the top three architectural styles by how they're compatible with the tools you have identified. With this list in hand, let's go to the next step.

Having a short list of architectural styles is helpful because you'll be filtering out the ones that aren't compatible with each of the restrictions you're introducing. Now is the time to examine your business objectives and understand which of the five architectural styles can help you reach your goals. To do that, let's return to the architectural style comparison matrix from *Chapter 5*. Now, identify the top attributes that are important to your business, and compare the short list of styles against them. As an example, let's see how REST, gRPC, and GraphQL compare against flexibility, cost, reliability, and interoperability. Note that this is just an example, and you should carefully select the attributes that matter the most to your business:

	REST	GraphQL	gRPC
Flexible	✓	✓	✓
Cost-effective	✓		
Reliable	✓	✓	✓
Interoperable	✓	✓	

Flexibility is common across all three API architectural styles. All the styles can adapt to changing business requirements and allow your organization to respond to any changes quickly. However, only REST can be genuinely considered cost-effective, since both GraphQL and gRPC generally have a higher cost of development. Looking at reliability, you can say that what all the styles provide is reasonable. All three API architectural styles can offer reliable communication that only depends on the quality of the implementation. Looking at interoperability, you can see that both REST and GraphQL are highly interoperable, while gRPC has its limitations. While both REST and GraphQL usually use JSON, which is a widespread format, gRPC uses its own Protocol Buffer binary format. Following this example, you can clearly see that gRPC isn't a good choice for your particular business needs. The next step is running the same exercise for technical constraints. Let's see how it goes.

Unlike business objectives, technical constraints refer to technology rules that can't be broken. Organizations have technical constraints to promote consistency across different projects and make it easier to manage the technology landscape. Let's continue our exercise and see how REST and GraphQL compare against a set of technical constraints. As an example, consider rate limiting, response caching, and versioning as the constraints to examine:

	REST	GraphQL
Rate limiting	✓	✓
Caching	✓	
Versioning	✓	✓

It is clear now that REST is the option to choose. While GraphQL can also offer rate limiting, its implementation tends to be complex because of the way it handles queries. With REST, you can easily add rate limits at the endpoint level. A similar situation happens with response caching. While REST offers HTTP caching ready to use, with GraphQL, you have to implement it yourself or follow a customized solution. Finally, versioning is not available in GraphQL because it promotes continuous improvements, through the deprecation of fields that aren't used. Overall, the choice of REST seems to be solid because it meets the needs of users, business objectives, and also, an organization's technical constraints.

Different types of APIs

You've been seeing examples of how certain API architectural styles compare against a few popular attributes. Let's now look at a more comprehensible list of API types to see what they look like. One way to split APIs into groups is to start with the ones that follow a synchronous communication and then look at the asynchronous ones.

REST

REST is probably the most popular API type in 2023. However, sometimes generic HTTP APIs are mistaken as REST because they share some of its attributes. Let's disambiguate and only consider REST APIs to be those that follow a set of characteristics and constraints:

- **Stateless**: Each request a client makes has to contain all the information the server needs. In other words, there's no notion of state being shared between requests.

- **Client-server architecture**: Clients and servers are separate entities. Clients initiate communication with servers by performing a request. Servers reply to requests by sharing responses with clients.

- **Resource-based**: The interface is based on resources that are exposed to clients. Resources are represented by **Uniform Resource Identifiers (URIs)**.

- **CRUD operations**: Resources can be accessed and manipulated through a set of common operations. **CRUD** stands for **Create, Read, Update, and Delete**. These are the operations that you can perform on most REST resources.

- **Hyperlinks in responses**: API servers can add hyperlinks to responses to help clients navigate between resources. This is called **Hypermedia as the Engine of Application State (HATEOAS)**.

- **Layered**: REST APIs can be decomposed into several layers that can exist independently. Examples of layers include authentication, load balancing, and caching.

- **Caching**: API servers can natively support HTTP caching. This method lets the caching happen on any of the intermediary layers of the API.

gRPC

Another synchronous API type that is popular in usage is gRPC. Unlike REST, gRPC architecture is based on the **Remote Procedure Call (RPC)**. By following this paradigm, it lets API clients execute operations remotely. gRPC also has particular attributes and constraints:

- **Independence**: gRPC servers and clients can be implemented and executed in a multitude of programming languages and frameworks. On top of that, it can generate server and client code from its **Interface Definition Language (IDL)**.

- **Performance**: Everything in gRPC has been designed with performance in mind. Its use of a binary serialization format makes communication less verbose and more efficient. Additionally, it uses the HTTP/2 protocol, which is considered more performant than its first version.

- **Bidirectional streaming**: In addition to offering one-way communication between clients and servers, every gRPC connection can also share information both ways. This attribute lets you build streaming features where data can flow between servers and clients easily.

- **Strongly typed**: The interface of gRPC servers offers a strict set of rules for how data and operations should be handled. Whenever clients and servers interact, they must adhere to the strongly typed interface requirements. In essence, strong typing helps prevent mistakes and makes APIs more reliable and developer-friendly.

GraphQL

As you can see, gRPC's attributes and constraints differ greatly from REST's. While REST is more focused on adhering to HTTP and being easy to use, gRPC's focus is on performance and efficiency. Let's now see what GraphQL looks like by looking at its most essential attributes and constraints:

- **Declarative**: Clients can obtain precisely the data they need by defining a query in a declarative way. In other words, as a client, you can tell a server which fields you want to retrieve.

- **Single endpoint**: With GraphQL APIs, you can perform any query using a single HTTP endpoint or URL. There are no separate URLs for different operations or resources. Instead, you perform queries on a single URL.

- **Hierarchical**: GraphQL APIs support queries that specify fields nested in hierarchical structures. You can, in a single query, obtain a whole tree structure of information.

- **Support for subscriptions**: Clients can subscribe to specific events, including data changes, and receive updates from the server. You can use GraphQL subscriptions to build chat applications and collaborative editing solutions, among other things.

- **Strongly typed**: GraphQL APIs use a strongly typed schema to define data structures and operations. Strong typing offers data validation, auto-completion, and documentation of queries, making the developer experience better.

It's interesting to note how different GraphQL is from gRPC and REST. Strong typing is the only attribute that it has in common with gRPC. Being different is not necessarily a bad thing, as it offers more diversity of choice. SOAP is another API type worth exploring and investigating. Let's see what SOAP's attributes and constraints are.

- **XML-based messaging**: SOAP APIs use XML as the data format to communicate information between servers and clients.

- **Strict contract definition**: API contracts are defined using the **Web Services Description Language (WSDL)**. You can describe the API's operations, parameters, and data types. Having a strict contract makes SOAP servers and clients agree on a data format and operational semantics.

- **Independence**: SOAP can be used on any network protocol, including HTTP and SMTP. The choice of network protocol can determine whether a SOAP API is synchronous or asynchronous. Additionally, you can implement SOAP servers and clients using a variety of programming languages and frameworks.

- **Extensible**: SOAP can be extended with features such as encryption, digital signatures, reliable messaging, and transactions, among others.

- **Formal error handling**: With SOAP, you have a standard way of describing and handling errors called faults. You can describe the code of the fault, a description of the error, and optional detailed information.

AMQP

Up until now, we've looked at synchronous API types. Well, not really – SOAP can be both synchronous and asynchronous, depending on the chosen network protocol. Let's get into the realm of asynchronous APIs and start by exploring the attributes and constraints of AMQP:

- **Messaging protocol**: AMQP enables the exchange of messages in distributed systems. With AMQP, you can reliably send and receive messages asynchronously.

- **Queueing and routing**: Messages can be queued before they're delivered to one or multiple consumers. AMQP offers routing capabilities that let you send messages to specific queues, based on rules.

- **Broker-based architecture**: Message brokers are intermediaries between producers and consumers. Brokers receive messages from producers, queue them, and then deliver them to consumers.

- **Multiple messaging patterns**: AMQP supports the point-to-point messaging pattern, where a message is delivered to a specific consumer; the publish-subscribe pattern, where messages are delivered to subscribing consumers; and the request-reply pattern, where producers send messages and expect a reply.

- **Reliable**: With AMQP, you can be sure that messages are received by consumers. The built-in acknowledgment system lets consumers signal that they've received a message.

- **Independent**: AMQP is language- and platform-agnostic, working with a variety of programming languages and systems. Additionally, it integrates well with other messaging systems.

- **Scalable and performant**: With features such as message batching, compression, and asynchronous processing, AMQP is considered highly scalable and performant.

MQTT

As you can see, AMQP looks like a good choice in environments where reliability, queueing, routing, and different message patterns are required. Let's look at another asynchronous API type, where the focus is on being lightweight and consuming low amounts of power. These are MQTT's most important attributes.

- **Lightweight**: By design, MQTT is a lightweight protocol that is meant to spend as little energy as possible. MQTT is well-suited for IoT environments and low-power devices.

- **Publish-subscribe messaging**: With MQTT, publishers send messages to a central broker that makes them available to subscribers. Messaging is totally asynchronous and dependent on the broker.

- **Quality of service**: MQTT offers the ability to specify levels of **Quality of Service (QoS)**. The supported levels are QoS 0, 1, and 2:

- **QoS 0**: Also called "at most once," this guarantees that messages are delivered to subscribers at most once. QoS 0 is often used in scenarios where a guarantee of delivery is not needed. One example is the fire-and-forget scenario, where messages are simply sent to the broker without any expectation of delivery to subscribers.

- **QoS 1**: Known as "at least once," this offers the guarantee that the delivery of a message happens at least once. This level is appropriate for situations where you want to make sure that subscribers receive your messages. It might involve message retransmission and end up resulting in duplicate deliveries.

- **QoS 2**: The "exactly once" level guarantees that messages are delivered to subscribers at least once. With this quality level, you can be sure that your messages are received by subscribers only once. It's used in situations where both message loss or duplication could have consequences.

- **Persistent sessions**: With MQTT, the broker can store and resume your consumer state whenever you connect. This allows the broker to have sessions that span multiple connections and retain messages for later delivery.

- **High throughput**: Through horizontal scalability, MQTT is designed to handle high amounts of messages. It can handle thousands of simultaneous connections and deliver messages to large numbers of subscribers.

- **Low latency**: Communication across MQTT participants has low latency. It lets you use MQTT in environments where real-time communication and quick response times matter.

At this point, you've seen what the different types of APIs look like. You now understand what each API type is best for and the elements that help you decide which one to use. Let's now look at ways to define APIs so that both people and machines can understand those definitions.

API specification formats

When defining an API, it's important to document the definition in a way that people can understand. However, that's not enough. API definition documents should also be understandable by code. In the end, API definition documents are meant to be interpreted and used by the software. API servers and brokers can use definition documents to programmatically make the features available to consumers and subscribers. On the other hand, consumers and subscribers can use the same documents to interact with the API producers. API definition documents are the glue between producers and consumers. There are different types of API specification formats that work well with APIs of different types. Let's look at the most popular specification formats to understand what they are and what types of APIs you can define with them.

OpenAPI

This is probably the most popular API specification format. OpenAPI started with a different name. It was created in 2010 with the name of Swagger. In 2015, its original creators donated Swagger to the Linux Foundation, which renamed it to OpenAPI. Since then, OpenAPI has evolved, and the two major versions in use today are versions 2 and 3. OpenAPI is a fundamental element in the definition of REST APIs, since it offers a machine-readable format – using either JSON or YAML – that can be used to generate documentation and API client and server code, and it works with a wide range of programming languages and frameworks.

You saw an example of what an OpenAPI definition document looks like in *Chapter 5*. The example was written in YAML and defined a payments API. Later in this chapter, you'll take that example and rewrite it in JSON to see what that's like.

IDL (protocol buffers)

IDL has been around for a while. It was developed by the **Object Management Group** (**OMG**) to define the structure and behavior of interfaces. With IDL, you can define the operations, data types, and communication protocols that software components use. One of the ways you can use IDL is to define how gRPC APIs behave. gRPC uses protocol buffers as its way of sharing data between servers and clients. Since you can generate code from protocol buffers, let's see how you can use an IDL document to define your protocol buffers.

Protocol buffer IDL documents start by identifying the type of syntax they refer to. In this case, protocol buffers are defined by the `proto3` identifier:

```
syntax = "proto3";
```

Then, you can define services and messages, among other things. A service is an identifier of a gRPC operation, where you define its input parameters and response:

```
service GreetingService {
  rpc SayHello(HelloRequest) returns (HelloResponse) {}
}
```

A message identifies a data type and its attributes. Messages can be used as an input parameter or a response to services. You can define the data type for each attribute and its position in the list of attributes. In our case, let's define the `HelloRequest` and `HelloResponse` messages, where both have a single string attribute called name:

```
message HelloRequest {
   string name = 1;
}


message HelloResponse {
   string name = 1;
}
```

If you haven't figured it out by now, this example gRPC API lets clients execute `GreetingService` by sending a name described by `HelloRequest` and receiving back a name, presumably the same one, described by `HelloResponse`.

The next step would be to compile this definition using a protocol buffer compiler. From there, you could generate client and server code. Having ready-to-use server code in your programming language of choice lets you or your development team focus on implementing the business logic behind each operation, instead of spending most of the time implementing the interface layer. Another API type that offers similar levels of automation is GraphQL. Keep reading to see how you can create a definition.

GraphQL

In 2015, Facebook released GraphQL publicly. While it was initially designed to address the limitations of other API architectural styles, it only gained popularity after it was made public – so much so that, in 2018, it drove the creation of the GraphQL Foundation, a collaboration between several companies, including Facebook and GitHub. The main advantages of GraphQL are its efficient and flexible approach to data fetching and the number of tools and frameworks that support it. It also offers an easy way to define your API by using a GraphQL schema.

You start by defining the data types and fields that are a part of your GraphQL data model. You do that by using the GraphQL **Schema Definition Language** (**SDL**), a format that serves as a human-readable and language-agnostic way to define the structure of an API. Similarly to gRPC protocol buffers, the SDL can be used to generate server and client code. Here's a straightforward example of defining a GraphQL API that lets clients send a `hello` query with name on it and receive back a salutation:

```
type Query {
   hello(name: String!): String
}
```

The SDL first defines a data type called `Query` with the single `hello` field, which receives a mandatory parameter called name. The type of the parameter is `String`, and the exclamation point at the end makes it mandatory. Finally, the output of the `hello` field is itself `String`.

This is a very simple example, just to illustrate how you can define a GraphQL API. There are other things you can define with the GraphQL SDL:

- **Type definitions**: You can define custom types that represent the data elements in your API. You can choose from objects, scalars, enumerations, and inputs. While objects represent the entities present in your data model, scalars represent primitive values, such as integers and strings. Enumerations represent a set of fixed values, and inputs define how input data for mutations or complex queries look.

- **Fields and relationships**: Inside each type, you can define fields that represent the properties of entities of the defined type. In the example you saw earlier, the `hello` field is a property of the `Query` type. Fields can also have parameters, as you saw with the `name` parameter of the `hello` field in the same example.

- **Query and mutation types**: You can explicitly define the `Query` and `Mutation` types. While the `Query` type lets you define operations that fetch data from your API, with the `Mutation` type, you can define actions that modify information.

- **Directives**: You can add custom behavior to queries and mutations. With directives, you can define behaviors such as data validation rules, authentication and authorization, and caching. There's a special type of directive called `@deprecated` that lets you mark fields as deprecated. Here's an example of how a caching directive would be defined:

```
directive @cacheControl(maxAge: Int!) on FIELD_DEFINITION

type Query {
  hello(name: String!): String @cacheControl(maxAge: 60)
}
```

After you have the SDL document finished, it's time to convert it into server and client code. GraphQL interfaces with business logic through functions, which are called every time there's a request for a query or mutation. These functions are called resolvers and are responsible for implementing the behavior of the API.

WSDL

The goal of WSDL is to describe the functionalities and interfaces of web services. And I'm using web services and not APIs on purpose. WSDL was created in the early 2000s when web services were extremely popular. Something else also trendy back then was XML. Because of that, WSDL uses XML as a standard way to define operations, message formats, and network protocols. WSDL is still relevant today, as it's the specification format used with SOAP APIs. WSDL documents let you define different elements of your API that are important to its operation:

- **Types**: You can define all the data types used in your API. You can define primitive data types such as integers and strings and also complex ones composed of other types.

- **Messages**: In SOAP, messages represent the information being exchanged between a client and a server. You can define the structure of a message and whether it's an input or an output of an operation.

- **Port types**: A group of related operations is called a port type. It consists of one or more operation definitions and acts as a blueprint for the API, guiding clients to construct and send appropriate messages to each operation.

- **Bindings**: Because SOAP can work in different communication protocols, you have to identify them. Each binding lets you define a protocol and the message formats that are used. Examples of bindings include SOAP over the HTTP protocol and SOAP over SMTP.

- **Services**: To define the endpoint where the API is located, you define a service. You can also identify the communication protocol and any protocol-specific elements relevant to operating the API.

Let's see what a simple WSDL document looks like. At first glance, it looks like HTML, the format used to build web pages, because it uses tags to define elements and attributes to define parameters. WSDL is, in fact, an XML document with some specific elements and attributes. You start by identifying that the XML document uses the WSDL format. You do that by creating the XML `definitions` tag at the very top of the document:

```
<definitions
xmlns="http://schemas.xmlsoap.org/wsdl/"
xmlns:soap="http://schemas.xmlsoap.org/wsdl/soap/"
targetNamespace="http://example.com/hello-service">
```

The `definitions` tag has two `xmlns` attributes. Those are meant to identify the XML namespaces of the document. The first one identifies the generic WSDL namespace, and the second one specifies the SOAP namespace. The third attribute, `targetNamespace`, identifies that the document defines an API with the specified value. Now, inside the `definitions` tag, you define the types you'll use in the API:

```
<types>
  <schema xmlns="http://www.w3.org/2001/XMLSchema"
targetNamespace="http://example.com/hello-service">
    <element name="helloRequest">
      <complexType>
        <sequence>
          <element name="name" type="string"/
        </sequence>
      </complexType>
    </element>
    <element name="helloResponse">
      <complexType>
        <sequence>
```

```
            <element name="message" type="string"/>
          </sequence>
        </complexType>
      </element>
    </schema>
  </types>
```

The data types are defined using XML schema inside the `types` element. XML schema is a specification that defines the structure, data types, and constraints for XML documents, ensuring consistency and validity. You can see that the schema defines two elements, the API request, and the response. Those will be used by the code running both the API server and the client. After having data types defined, it's time to identify the request and response messages. The definition is meant to relate the messages to the data types:

```
<message name="helloRequestMessage">
  <part name="parameters" element="tns:helloRequest"/>
</message>
<message name="helloResponseMessage">
  <part name="result" element="tns:helloResponse"/>
</message>
```

You can see that there are two identified messages. The first one associates the request data type previously defined with the request message. The second one does the same but for the response data type and message. You're now ready to define the port types, which identify what operations the API offers to clients:

```
<portType name="helloPortType">
  <operation name="helloOperation">
    <input message="tns:helloRequestMessage"/>
    <output message="tns:helloResponseMessage"/>
  </operation>
</portType>
```

There's one single port type with a single operation. The definition identifies an input associated with the request message and its corresponding data type, and an output associated with the response message and its data type. The next step is identifying the binding. In other words, you define how the API will make the operation available to clients:

```
<binding name="helloBinding" type="tns:helloPortType">
  <soap:binding style="document" transport="http://schemas.xmlsoap.
org/soap/http"/>
  <operation name="helloOperation">
    <soap:operation soapAction="http://example.com/hello-service/
hello"/>
    <input>
```

```
      <soap:body use="literal"/>
        </input>
    <output>
      <soap:body use="literal"/>
    </output>
  </operation>
</binding>
```

The binding definition starts by identifying the port type and the protocol used for communication. In this case, the API will communicate using documents over HTTP. Then, it associates the operations with an HTTP endpoint and defines what type of input and output bodies the API will have. Finally, after learning so many details about the API, it's time to define the service as a whole:

```
<service name="helloService">
  <port name="helloPort" binding="tns:helloBinding">
    <soap:address location="http://example.com/hello-service"/>
  </port>
</service>
```

Here, you can see that the service definition is a mere impression of what had been previously defined. The service is given a name and is associated with an existing port, with a specific binding. Finally, the SOAP HTTP root address is identified. With this information, you have everything you need to define a SOAP API using WSDL… well, not quite everything. Last but not least, before we forget, you should close the `definitions` XML tag that you opened at the very beginning of the WSDL document:

```
</definitions>
```

AsyncAPI

AsyncAPI is an open specification that provides a standardized way to describe event-driven APIs, enabling the documentation, design, and management of asynchronous communication patterns by defining the format, message types, endpoints, and protocols used to exchange messages, facilitating interoperability and understanding between producers and consumers of event-based systems.

With AsyncAPI, you can, for example, define APIs that use AMQP and MQTT, two technologies covered earlier in this chapter. That means that AsyncAPI is not tied to any particular technology or protocol. Instead, you can use it to define the communication patterns you want to support in your asynchronous API. You use AsyncAPI by creating an API definition document, using either the JSON or YAML format. Since we covered the basics of AsyncAPI in *Chapter 5*, let's now see in detail what you can do with the `channels` and `components` sections.

The AsyncAPI `channels` section lets you identify how messages are published and subscribed. Each channel corresponds to a publishing destination that API clients can use to make their messages available to subscribers. You can consider channels as message topics where events are exchanged. Each channel has its own identifier and lets you specify whether clients can publish it, subscribe to it, or both. You can define the following attributes for each channel operation type – publish or subscribe – that you create:

- **Operation identifier**: The identifier of the operation you're defining. This should be a unique identifier within the whole AsyncAPI definition.

- **Message**: The message schema associated with the operation. Knowing the message schema is valuable because it lets publishers and subscribers validate messages reliably.

- **Summary**: You can identify a short summary of what the operation does. In generated documentation, the summary can be used as the human-readable title of the operation.

- **Description**: A lengthy description of the operation. This can be used to provide detailed instructions on how to use the operation.

- **Bindings**: This is how you define which protocols or broker-specific features the operation will work on. You can identify bindings specific to protocols such as AMQP or MQTT.

In the following section, you'll see how the different API specification formats that you've learned about can be used to define the web payments API example that we've used throughout the book. Keep reading to get hands-on with machine-readable API definitions.

Creating a machine-readable API definition

Even if creating a machine-readable API definition sounds like a daunting task, its benefits are worth the effort. Let's see some of the things you can get after you have an available machine-readable API definition:

- **Automation and tooling**: From automated client and server code generation to generating documentation, there are many tools available, offering varied degrees of automation, to convert a machine-readable definition into other assets.

- **Interoperability**: Having a machine-readable API definition lets you easily move from one programming language or framework to another. More importantly, if you can do this, so can your API users. They can easily consume your API using their favorite programming language.

- **Discoverability**: Machine-readable definitions can be used to generate interactive documentation and sandboxes, where potential users can explore and sign up to use your API. This increases the discoverability of your API product, as it makes it easy to understand what it does.

- **Contract enforcement**: Having a machine-readable definition is like having a contract, enforcing rules on how API consumers can interact with your API and how a server will respond to their requests. The machine-readable definition is the source of truth.

- **Versioning**: With a text-based machine-readable API definition, you can easily understand whenever changes are made. Tracking changes to the definition makes it easier to manage potential breaking integrations and proactively communicate with your customers.

- **API governance**: If consistency and standardization are things that you're interested in, then having a machine-readable definition is a must. Anything from naming conventions to error messages can be checked against governance rules.

- Now that you know about the advantages of having a machine-readable API definition, let's create one in each of the specification formats that we're studying. To make things easier, I created machine-readable documents for each of the specification formats you've been reading about. The machine-readable API definition documents are meant to be an example of what you can do. Please see them as a reference that you can access at any time during your journey of building an API product. The following documents are available on the `PacktPublishing/Building-an-API-Product` GitHub repository in the `Chapter08` folder:

  - `example-openapi.json` and `example-openapi.yaml`: An OpenAPI example definition using the JSON and YAML file formats, respectively

  - `example.proto`: An example Protocol Buffers definition

  - `example.graphql`: An example GraphQL definition

  - `example.wsdl`: An example WSDL definition

  - `example-asyncapi.json` and `example-asyncapi.yaml`: An AsyncAPI example definition using the JSON and YAML file formats, respectively

Following API governance rules

The primary goal of API governance is to align your API with your business objectives. It does that by offering a framework of processes, policies, and practices that help you adhere to industry standards and ultimately meet users' needs. The API governance framework often interacts with various steps of the API life cycle. API design is one of those areas that is governed by a set of guidelines. Other areas include versioning, security, documentation, performance, and API life cycle management. Overall, the goal is to have control over all API-related operations and activities. Let's see how some of the areas are influenced by and benefit from an API governance strategy.

API design

By following the steps you learned in *Chapter 5*, you can make sure that you design your API product in a way that is aligned with what users need. By following an API governance strategy, you can also ensure that user alignment isn't creating an unnecessary burden on your business goals.

The first thing that people think about when building or consuming an API is how consistent it is with other APIs. **Consistency** is a key factor in consuming and maintaining an API. Even if you only have one API, it's important that it's consistent with other APIs in your sector, especially the other APIs that your users consume regularly. API governance promotes a set of common design patterns, naming conventions, and data formats for you to follow. The main point is that you don't have to reinvent every request and response, error message, authentication of your API, and so on. Instead, you'll use well-known standards that can be internal to your company or already established in the industry.

An excellent example of one of these standards is the problem details for HTTP APIs, described by the Internet Engineering Task Force RFC 7807. This **Request for Comments (RFC)** defines a standardized way of conveying information about HTTP errors, helping consumers obtain meaningful information whenever a problem occurs. You can come up with your own way of describing errors. However, if you follow what RFC 7807 proposes, you can be sure that your API will be consistent with others that follow it, and this consistency makes your API easier to use.

The second area where API governance helps is the ease of use. Firstly, it encourages the modularity and reusability of operations, data types, and other elements of your API. Secondly, it promotes the API user experience by encouraging the use of well-defined interfaces that consumers can quickly and easily understand. Reusing elements of your API as much as possible increases consistency. If two operations have a similar input, attempting to reuse it makes sense. If you're using identical error codes and messages across different operations, being able to standardize them makes sense. Additionally, offering an API that users can grasp without the need to consult documentation can dramatically improve its usability.

Not having to consult documentation comes at the expense of following naming conventions. By naming things in a way that API consumers already understand, you're decreasing the cognitive load of interacting with your API. Here are examples of some of the most popular naming conventions across different types of APIs:

- **Plural nouns to represent REST resources**: The words used in the URLs of a REST API are plural nouns because they represent collections of resources –for example, `/payments`.

- **Use of HTTP verbs to access REST endpoints**: HTTP offers standard ways of interacting with web resources. In a REST API, you use verbs to indicate the type of action that you want to execute – for example, `POST /payments` to create a payment and `GET /payments` to retrieve a list of payments.

- **Use of identifiers to access individual REST resources**: While accessing a collection of resources is useful, being able to retrieve, update, or delete a single resource can be handy. REST APIs let you access individual resources by using one of the HTTP verbs, in combination with a URL that identifies the resource with an identifier – for example, `GET /payments/28e8203b-d857-4048-9ddd-7f6489f6794e` to get a single payment and `DELETE /payments/28e8203b-d857-4048-9ddd-7f6489f6794e` to remove it.

- **Singular nouns to represent data types in GraphQL**: It's common practice to use singular names for types. Instead of defining a payments data type with the word `Payments`, which is plural, you'd use the singular form, `Payment`.

- **Use of verbs for GraphQL mutations**: It's common to use verbs to define GraphQL mutations to let users understand the action being performed and the affected data type – for example, `createPayment`.

- **Using the word service to identify gRPC services**: In gRPC, you identify a service by appending the word `Service` to the end of the name of the functionality – for example, `PaymentService`.

- **Identifying gRPC request and response messages with specific words**: You identify gRPC request messages by appending the word `Request` to the method name – for example, `CreatePaymentRequest`. Similarly, you identify response messages by appending the word `Response`.

- **Noun-verb structure to define event names in asynchronous APIs**: The noun-verb structure comprises a noun followed by a verb in the past tense. You use the noun-verb structure to identify event names – for example, `PaymentProcessed`.

These are just some of the conventions that people have adopted for different API types. By following these conventions, you're making your API consistent with what exists in the industry and the lives of consumers easier. Nevertheless, even if you follow all the conventions, there are times when the best way to guide users is to have concise documentation. The role of API governance in the influence of documentation is that it provides rules and a style guide that you and your team can follow when writing user-facing documentation. Again, the purpose is to have consistency. In this case, you want to have consistent documentation and follow the same style that you already use across your company. By presenting documentation in a way that users are already familiar with, you're improving their ability to understand the API. Documentation is one of the pillars of good API design and also a part of another area that API governance influences. Keep reading to learn how API life cycle management can gain from following the direction of API governance.

API life cycle management

With API life cycle management, you have a thread that guides you through the different stages of your API. All the activities that contribute to the design, development, deployment, operation, and evolution of an API can be managed and governed. The goal now is to obtain consistency in these processes to achieve a high degree of optimization and organizational structure.

Let's see how API governance can influence the consistency of the way you manage your API life cycle. The first thing API governance provides is a way to follow industry best practices and standards for each of the steps of the life cycle. That's important, as you don't want to spend time reinventing how to manage and operate your API when someone else already did it for you.

Another area that substantially improves from following the guidance of API governance is what happens every time you need to make a decision. By setting policies for decision-making, you'll have your activities documented, and you'll know what to do in every situation. Processes such as API design reviews can have playbooks that you and your team can follow. API security assessment sessions can be fully documented. API versioning can be done with confidence by following pre-established rules. Any activity surrounding your API becomes part of a bigger picture that is fully documented and can be easily replicated.

One final area that I want to highlight is team collaboration and stakeholder management. This is an area that is often neglected by API practitioners. However, without cooperation between all API stakeholders, it's very complicated to build an API product. API governance encourages collaboration among all the stakeholders, including customers, developers, software architects, business owners, operations teams, and security experts. By offering tools and guidelines for collaboration, you'll have a better-engaged team of stakeholders that can confidently participate in the decision-making process of building an API product.

Summary

Now, you know that choosing the right API architectural style involves a combination of understanding user personas, business objectives, and technical constraints. You also know how to create a machine-readable definition document for the type of API that you decide to use. Finally, you know the importance of API governance and how applying it to your design and life cycle management processes helps improve your chances of success.

When you began this chapter, you already knew how to define and validate your API design. You took what you had learned and started applying it to creating something concrete, in the shape of a machine-readable API definition. You learned that to pick the right architectural style, you need to connect what you learned from user personas with your business objectives, as well as your technical constraints. Those three criteria help you to evaluate the feasibility of all available API architectural styles and choose the right one. After understanding the process of choosing the architectural style, you learned about the different types of APIs. You became aware of the differences between API types, including synchronous and asynchronous APIs. You learned specific details about each API type, helping you cement your ability to choose the right one for your API product. Then, you got familiar with examples of machine-readable API definition documents, using OpenAPI, protocol buffers, GraphQL schemas, WSDL, and AsyncAPI. The same example API was used for all the definitions so that you could see the differences. After that, you learned about API governance and how important it is to improve the design of your API. You understood that following naming conventions and using well-known standards has numerous advantages to you and your potential API consumers. Finally, you learned the importance of applying governance principles to your API life cycle management. By standardizing the practice of life cycle management, you don't have to reinvent every step, and you will gain productivity and consistency.

Here are some of the things that you learned in this chapter:

- **The most popular API architectural styles**: REST, gRPC, GraphQL, SOAP, AMQP, and MQTT
- The connection between user personas, business objectives, and technical constraints in choosing an architectural style
- **The most common API specification formats**: OpenAPI, IDL, GraphQL schemas, WSDL, and AsyncAPI
- The most important characteristics of common API specification formats
- How some API specification formats can be used to define different types of APIs
- How the same API is defined using different API specification formats
- The meaning of API governance in the process of designing and managing an API
- How to apply API governance rules to your API definition
- What naming conventions are and why they're important
- The importance of API governance in life cycle management

Now, you're ready to start implementing your API product. You know the most critical elements of API design, the importance of having a sound API strategy, how to define and validate your assumptions, and finally, how to create a machine-readable API definition that you and your team can use during the implementation stage. Let's now implement the API. Keep reading to see how.

Part 3:
Implementing an API Product

Part 3 serves as a comprehensive guide to various aspects of API development, starting with a beginner-friendly approach covering language selection, code generation, prototyping, and business logic extension. It then delves into API security, emphasizing authentication, authorization distinctions, and introducing the security testing technique of fuzzing. This part further explores API testing methods, including contract testing, performance testing, and acceptance testing aligned with user personas. It concludes by covering API Quality Assurance, introducing behavioral testing for behavior validation and setting up periodic API monitors.

In this part, you'll find the following chapters:

- *Chapter 9, Development Techniques*
- *Chapter 10, API Security*
- *Chapter 11, API Testing*
- *Chapter 12, API Quality Assurance*

9
Development Techniques

API development is as complicated as the tools you use to make it happen. By following the steps presented in this chapter, you'll be able to use the right tools for different stages of development. You'll learn about the differences between mocking and prototyping and why prototyping is closer to having a final API server running. You'll also understand how you can choose the right programming language and framework for your API product. With that information, you'll be able to generate running API server code from an existing machine-readable definition document.

We'll start this chapter by recalling what we've already learned about API mocking. You'll understand that mocking is mostly used for validating an API design.

Once the design has been validated, you can create a prototype of your API server. You'll understand that the goal of prototyping is to have something running quickly. And to achieve that, you have to leave the implementation of the business logic for a later stage. You'll learn that choosing the most appropriate programming language and framework can be done by following a simple method. You'll see all the factors that can influence your decision. You'll then get a high-level overview of the most popular programming languages and their frameworks. Finally, you'll learn how to generate your API server code using two different approaches.

By the end of this chapter, you'll be able to take a machine-readable API definition document and transform it into running code. You will know how to do that using the programming language and framework best fit for your business requirements. You will also know that you can measure factors such as the learning curve, community support, job market, and performance to influence your decision. In the end, you will have specific knowledge of popular programming languages such as Node.js, Python, and Ruby. Finally, you'll have a running API server prototype that you can generate as many times as you want using any of the available programming language and framework combinations.

Here are the topics that you'll learn about in this chapter:

- Prototyping an API
- Choosing a programming language and framework
- Generating server code from a specification

Technical requirements

In this chapter, you'll interact with tools and systems that generate code from an existing machine-readable definition document. To benefit from this, you should be familiar with JSON and JavaScript and be able to execute commands on the terminal.

Prototyping an API

In previous chapters, you learned how to use an API mock server to validate the design of your API. Now, you'll learn how to use an evolved technique that's the first step in the API implementation. But first, let's review what API mocking is so that you can see how it relates to this stage of development.

Mocking is often used as a way to simulate the behavior of an API. It's considered a low-effort technique that you can apply without writing any code. Because of that, it's applied during the API design stage when you're validating your assumptions with stakeholders. You create an API mock that returns simulated responses, usually involving fake data, and you share it with potential API users so that they can test it. The goal is to help stakeholders understand how the API would work in a real scenario.

Once the design of the API has been validated, you can keep the mock server running so that your development team has something to start with before implementing an API prototype.

And that's where we are now. We're at the stage when your API has been validated, you already know what architectural style to use, you have already applied the constraints of the API type of your choice, and you now have a machine-readable API definition document that uses the appropriate specification format. It's time to start implementing.

The objective of the prototype is to create the first version of your API. However, unlike mocking, you want the first version to follow all the discoveries you've made during your API journey. You want the prototype to use the machine-readable API definition and the architectural style of your choice. Also, unlike mocking, with a prototype, you'll work with real data – or, at least, with a subset of your real data. Your goal is to make the first version as close as possible to the final implementation without it being the full implementation. To do that, you'll have to make some compromises. One of the things you don't need to implement now is the full business logic behind every capability. Instead, you'll need to implement generic versions of what the business logic would do. For example, if the business logic for one of the capabilities can be done using off-the-shelf solutions, it's preferable to do so instead of building a fully customized version. Your goal is to prototype the interface with as little effort as possible.

For example, instead of implementing full user signup and authentication, you can use one of many existing solutions, such as Auth0 or Amazon Cognito. Instead of working on all the details of a payment implementation, you have solutions such as Stripe and Adyen. Notifications can be built using solutions such as Twilio and AWS SNS. You get the picture. There are multiple off-the-shelf solutions for almost every type of business logic you can think of, and it would be best if you took advantage of that while you're prototyping your API:

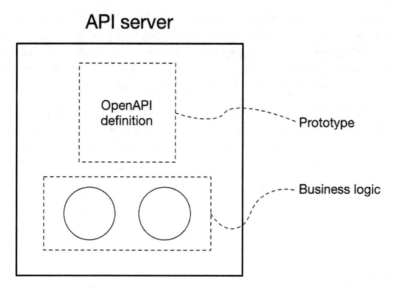

Figure 9.1 – Composition of an API server using a prototype and business logic

Additionally, you can obtain most of the mentioned solutions in a state that's ready to integrate with whatever code you have powering your API server. Most of the solutions are available for various programming languages and frameworks. This lets you choose the programming language and framework that best fits your needs without being constrained by a particular solution. Keep reading to learn what factors can influence your choice of programming language and framework.

Choosing a programming language and framework

Not being constrained by a particular choice of technology is great when you have other factors to consider. In any decision involving technology, you end up choosing based on a mix of business requirements.

Since the needs of users have already been guaranteed by the API design, you only need to make sure that you can implement what you designed using the programming language and framework of your choice. Assuming that, let's see what criteria you can follow during your decision process.

Factors to consider

The first thing to look at while choosing a programming language and framework is the list of requirements for your API product. These are often associated with the business objectives you identified in *Chapter 6* and the API design you studied in *Chapter 7*. The first requirement to look at is related to the ability to implement your API server in the architectural style that you have identified. While it's true that you can implement any architectural style using any programming language, some are more appropriate to your architecture than others. Keep reading to see a complete comparison of the most popular programming languages.

The second factor to consider while picking a programming language and framework is the learning curve. Choosing a language that takes a long time to master is counter-productive. The learning curve can be influenced by the programming language but also by your team's ability – and availability – to learn. Learning curves can be categorized according to their steepness. You start at the bottom, and you need to get to a point where you can use the language professionally. The more steep a learning curve is, the faster people acquire the knowledge necessary to get to the top of the curve.

The less steep a learning curve is, the longer it takes for someone to learn. The solution is to pick languages you and your team – or company – are already familiar with and have a learning curve that provides a progressive learning experience:

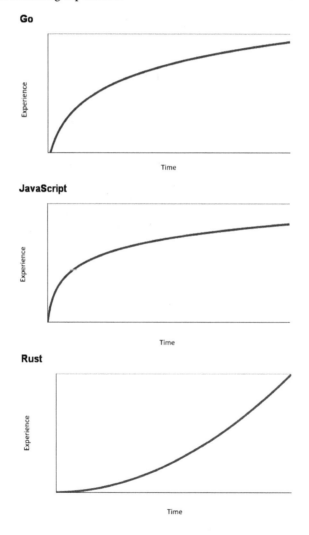

Figure 9.2 – Examples of learning curves for the Go, JavaScript, and Rust programming languages

Another factor that influences how people learn about a programming language is how much support they can get from the community. A community that is active and offers support often creates numerous learning resources. These can include blogs, tutorials, documentation, videos, and forums. With these resources, you can find guidance, tutorials, and solutions to common challenges related to the programming language that you're using. Additionally, a robust community is associated with software libraries and frameworks that can help you succeed by shortening the time it takes to develop your API. Another advantage of a strong community is the number of professionals that can be available to hire.

A programming language is only as good as those who can work with it. One factor to consider is the job market for the language that you're considering. You should look at the number of companies hiring, the level of experience that companies search for, and the number of people sharing their knowledge. You can use popular job boards and social networks that engineers use. On LinkedIn, a professional social network, you can search for jobs in a particular programming language and also for people with experience in that language. To be more thorough in your search, you can go to GitHub, a platform targeted at developers, where you can see the code they share. Ultimately, you want to gauge the supply and demand of professional with experience in the programming language you're choosing. One of the things that a demanding job market creates is an increase in the cost of hiring engineers.

Understanding how expensive your API development will become is critical to achieving success. As you've just seen, hiring can directly affect the cost of your API. The choice of programming language can also generate other costs. The most obvious cost is related to licenses. While many programming languages are fully open source and free, many frameworks have licenses prohibiting commercial use. Paying attention to the license is crucial if you wish to reduce your development costs.

Another sometimes unseen cost is associated with the performance impact of a programming language. Some programming languages and frameworks are less performant than others. If your choice of programming language introduces a performance penalty, then you'll have to increase the performance of your API using other methods. These include scaling your API horizontally by adding more servers, scaling vertically by increasing the memory and CPU of your API server, or adding layers of caching and pre-processing to your infrastructure. The more elements you add to the infrastructure, the higher its complexity becomes and the harder it is to maintain.

In summary, these are the most important factors you need to consider when you're choosing a programming language and framework to develop your API:

- **Requirements**: How the programming language can adapt to your business objectives and API design
- **Learning curve**: How difficult it is for someone to gain proficiency using the programming language and framework

- **Community**: How robust the community is around a programming language and framework and how its members collaborate
- **Job market**: What is the offer and demand for professionals with experience in the programming language and framework?
- **Cost**: What are the hiring and licensing costs of the programming language and framework?
- **Performance**: How performant the programming language and framework are in delivering the API you designed

Now, let's look at how well-known languages rate regarding these factors so that you have a starting point on what to choose.

Popular languages for building APIs

According to GitHub, a popular service for sharing code, the number of programming languages used in 2022 reached 500. However, not all languages are appropriate for building APIs. From those 500 languages, I've picked a short list of six that I consider the best for working with APIs. First, you can build backend code – code that runs on servers – with all these languages. Second, each is known for being widely used in at least one API architectural style. Let's start with a summary of each of the programming languages.

Node.js

Node.js is a JavaScript runtime environment created by Ryan Dahl in 2009. The promise is that whoever knows JavaScript can also write code that runs on the backend. Node.js grew in popularity partly because of its modular way of working with software libraries. **Node Package Manager**, or **npm**, has over one million modules that are easy to include in your code. npm is considered the world's largest software registry.

JavaScript

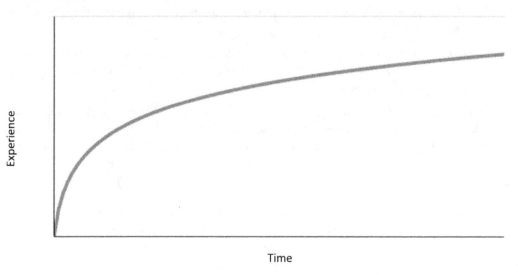

Figure 9.3 – Example of learning curve for JavaScript

Among these modules, there are a few that you can use to build an API:

- **Express**: A fast, minimalist web framework with over 28 million weekly downloads. It offers comprehensive documentation and extensive articles and tutorials shared by community members.

- **Koa**: An expressive web framework with a minimal footprint. Koa's strength is its middleware system, which lets you stack different actions between an HTTP request and a response. There's plenty of documentation and support available.

- **GraphQL.js**: This is the official implementation of GraphQL for Node.js and is directly supported by the GraphQL Foundation. It lets you build GraphQL APIs with minimal effort, letting you focus on the capabilities and abstracting all the architectural details.

Python

Guido van Rossum, a software engineer, created Python in the late 1980s as a language focused on simplicity and ease of use. Its current version, Python 3, was introduced in 2008, and it's a major revision that introduced breaking changes relative to version 2. Python is often used in teaching environments as an introductory way of learning software development. Python has a large community that helps with the development of the language and supports newcomers. Here are a few frameworks that are useful for building an API:

- **Django**: A high-level web framework that makes it easy to develop an API, it focuses on speed, security, and scalability. Django has its own community and is sponsored by the Django Software Foundation.

- **FastAPI**: A performant web framework for building APIs. As its name implies, it's a fast framework that also reduces the effort of developing an HTTP API. FastAPI has a thriving community powered by numerous sponsors.

- **Flask**: A microweb framework that doesn't require any tools or libraries. It's very simple and has almost no dependencies. However, it also doesn't offer common abstractions that other frameworks do. It has extensive documentation and works well with other Python modules.

Ruby

The goal behind the creation of Ruby was to combine all the best features of other popular languages while keeping it easy to use. That's what Yukihiro Matsumoto did in 1993. The first release of the language happened in 1995, and it's gained popularity since then. There was, however, a moment when it surged in popularity after the release of the now-popular Ruby on Rails framework. Ruby has a robust community of enthusiastic developers who create resources and offer support. Here are a few API-related frameworks:

- **Ruby on Rails**: This is probably the most popular Ruby web framework. It is a full-featured web framework that offers everything you need to get your API off the ground. There's an enthusiastic community ready to help you if all the available tutorials and documentation aren't enough.

- **Sinatra**: A low-effort framework for quick web development. It's a small and flexible framework with almost no dependencies. However, because of its simplicity, you have to build a lot of things by hand. It offers comprehensive documentation, though.

- **Padrino**: This is a web framework built on top of Sinatra that offers more features. While it keeps the simplicity of Sinatra, it provides more abstractions, making the development effort lower. Padrino is light and fast and can be used for projects of any size. You can easily find guides and tutorials on how to get started.

Java

James Goslin was working on a project to develop software for embedded devices in 1991 when he created Java. The language quickly transformed into a general-purpose tool and was announced to the public in 1995. Perhaps due to its type of commercial licensing, Java has proliferated in the enterprise world. The company behind Java, Oracle Corporation, offers different kinds of support and tooling. Additionally, Java has a strong community of enthusiasts that guarantee the language is well maintained.

The following are some well-known Java web frameworks:

- **Spring**: A complete programming model that includes different ways to get started with little or no coding required. Spring is a product of VMware Tanzu, which offers courses, certifications, and support.

- **Quarkus**: An open source Kubernetes-native Java framework designed for building lightweight and highly efficient Java applications. It's specifically optimized for APIs and cloud-native development, aiming to reduce startup time and memory consumption.

- **Grails**: An open source web framework that uses Groovy, a language built on top of Java. Its goal is to provide a high-productivity environment where most of the complexity is abstracted. It is widely used and supported by the Grails Foundation.

Go

Go, commonly known as Golang, was created by Robert Griesemer, Rob Pike, and Ken Thompson while working at Google. The language was developed internally in 2007 and publicly announced in 2009. The objective was to create a language that could be as efficient as other lower-level languages while being as easy to use as others at a higher level. Go is an open source language that's supported by a robust community.

Go

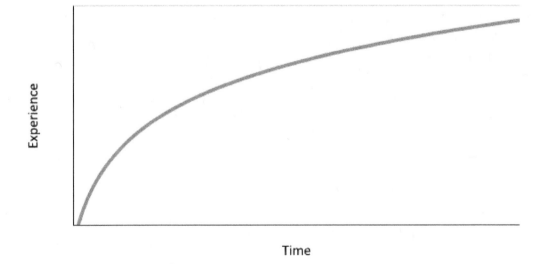

Figure 9.4 – Example of learning curve for Go

Here are some popular web frameworks:

- **Gin**: It's self-proclaimed as the fastest full-featured web framework for Go. It's built with performance and productivity in mind, supports a wide range of middleware plugins, and has built-in JSON validation. It offers comprehensive documentation and an energetic community.

- **Beego**: Used for building enterprise-scale REST APIs, it's inspired by Sinatra and Flask, among others. It offers sparse documentation. However, it's backed by a vibrant community.

- **Fiber**: A web framework inspired by Node.js Express. It's designed to make it very easy to develop performant APIs. Extreme performance is one of its selling points. It offers comprehensible documentation and a robust community.

Rust

The Rust programming language was developed as a personal project by Graydon Hoare while working at Mozilla in 2006. The language gained popularity when Mozilla officially announced it in 2010. Rust was built as a language that offers safety, concurrency, and speed. It has a package manager called Cargo and a robust community that involves various working groups who are responsible for different areas of the language.

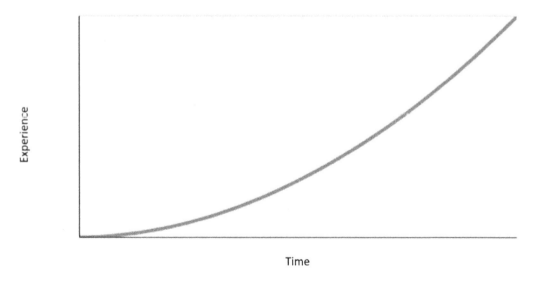

Figure 9.5 – Example of learning curve for Rust

The following are some well-known Rust frameworks that you can use to build your API:

- **Rocket**: Inspired by Ruby on Rails and Flask, among others, it's a performant, flexible, and easy-to-use framework. It's backed by a strong community and offers plenty of documentation.

- **Actix**: A powerful and pragmatic web framework built with performance in mind. It's feature-rich but is extensible as you can add your own libraries to the base framework. Different types of documentation are available, along with a vibrant community.

- **Axum**: A framework that focuses on modularity. It offers strong middleware and makes it easy to develop APIs. Its documentation is primarily a reference to the code itself. However, there's a community that's available to chat and offer support.

Comparing programming languages

By now, you should be able to compare programming languages using the factors that I mentioned or your business requirements. Let's look at how I would make the comparison so that you have a starting point to make your own decisions. The following table shows the popular languages on the *y axis* and the most important factors on the *x axis*:

	Learning Curve	Community	Job Market	Performance
Node.js	Gentle	High	High	Medium
Python	Gentle	High	High	Low
Ruby	Moderate	Medium	Medium	Low
Java	Moderate	High	High	High
Go	Moderate	Medium	Medium	High
Rust	Steep	Medium	Low	High

You can use this table to help you decide which language to use based on your business requirements. If you want to use a performant language that is relatively easy to learn and has a high job market, you would probably pick Java. If you're not too worried about performance but would like a programming language that is very easy to learn, then you'd choose Node.js. In the end, what's important is that you can use the programming you choose to build your API. Keep reading to see how you can generate the server code for your API.

Generating server code from a specification

In *Chapter 8*, you learned how to translate your API definition into a machine-readable document that can be used, among other things, to generate server and client code. You examined different specifications and had the chance to understand their differences. Now is the time to take one machine-readable document and convert it into code that runs and powers your API server. This code is meant to act as the prototype of your API because it doesn't include any business logic.

We'll focus on one of the specification formats and one of the programming languages. I'll use the OpenAPI specification format and the Node.js programming language to show you how you can quickly get your API server up and running.

Now, let's see how you can generate server code using two different approaches. The first approach uses Postman, a popular API platform, so that you can go from your OpenAPI definition document to a fully working API server. The second approach uses a well-known open source code generator called OpenAPI Generator to create the server code from the same OpenAPI definition document.

Generating server code using Postman

To begin with, go to Postman and either create a new account or sign in using an existing one. When inside Postman, create a new API by clicking on the **New** button from the left-hand pane and clicking **API**. This should open a screen where you can enter information about the API you just created. Let's skip that for now and go straight to **Definition**. Click on the plus (+) button and then **Author from scratch**. You'll be presented with a choice of the definition type – the specification format – and the format of the definition file. Choose **OpenAPI 3.0** as the definition type and **JSON** as the definition format. Finally, click on the **Create Definition** button.

You will see a screen with a blank definition editor where you can enter your machine-readable definition document. Copy the OpenAPI document from *Chapter 8* and paste it there. Click on the **Save** button; you'll see the screen update to show you the outline of the definition, including its paths and components. Now, open the main screen of the API again – you'll see a code button (</>) on the right-hand side pane. Click on it; you'll be taken to the **Code Generation** screen:

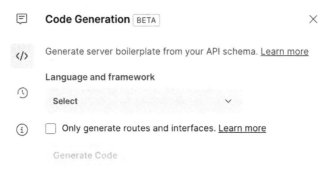

Figure 9.6 – The Code Generation screen

While on the **Code Generation** screen, pick **NodeJs - Express** from the **Language and framework** list of options and click on the **Generate Code** button. You will see that a ZIP file with the generated code is downloaded. Opening this ZIP file will reveal the contents of a Node.js fully working REST API server.

Here's the structure of the generated code:

- `package.json`: The Node.js project definition. Here, you can find the name and version of the project and any dependencies.

- `README.md`: A simple project document where you can find instructions on how to run the server.

- routes: A folder that contains any generated server routes. Routes are how the Express framework lets clients make requests to the server:

  - payments.route.js: The generated payments route. This route was generated from the /payments path that you defined in the machine-readable document.

- routes.js: The code responsible for resolving the requests and calling the right routes.

- server.js: The code responsible for starting the server and initializing the routes.

- services: A folder that contains any generated server services. This is where you would add your business logic:

  - payments.js: The generated business logic for the path. It includes all the information available in the machine-readable definition. However, it doesn't include the code to make the actual payment.

The server is ready to be started. You can do that by either running it on your computer or by uploading the code to a server on the web. I encourage you to try running the code on your own to see that it will start a prototype of your API. After the server is running, you can try it out by making a request and seeing the response.

Whenever you make an HTTP POST request, you will get a response with the following content:

```
{
    "transactionId": "<uuid>"
}
```

The actual ID of the transaction is not generated. If you inspect the file where the response is created, you'll see that there's no code generating an ID. To do that, open the payments.js file inside the services folder. You'll see the following code generating the response:

```
var data = {
    "transactionId": "<uuid>",
},
status = '201';
```

To add business logic, you'd have to expand this code to generate an ID and attempt to make a payment.

Generating server code using OpenAPI Generator

Using OpenAPI Generator requires medium technical knowledge. You'll have to install some scripts on the computer where you'll run the generator and sometimes, that can feel complicated. If you're on a Mac, the easiest way is to use Homebrew to install it (please visit `https://openapi-generator.tech/docs/installation/` for all the installation options). Homebrew is an application and package manager for Mac computers. If you don't have it, install it now by going to `brew.sh` and following the instructions.

To install OpenAPI Generator using Homebrew, run the following command on a Terminal:

```
brew install openapi-generator
```

After installing OpenAPI Generator, you can invoke it by running `openapi-generator` on a Terminal. To generate the server code for your OpenAPI definition document, first, create a file called `payments.json` and paste into it the OpenAPI machine-readable document from *Chapter 8*. Then, run the following command on a Terminal in the same folder where you have created the `payments.json` file:

```
openapi-generator generate -i ./payments.json
-g nodejs-express-server
```

This command tells OpenAPI Generator to generate server code in the Node.js Express framework using the definition from the `payments.json` file.

Here's the structure of the generated code:

- `api/openapi.yaml`: A copy of the OpenAPI definition you provide in YAML format.
- `config.js`: A set of configuration options that are used by the server.
- `controllers`: A folder containing the code responsible for translating the client requests into calls to server code.
- `expressServer.js`: The code responsible for running the API server.
- `index.js`: The main code responsible for starting the Node.js project.
- `logger.js`: The code responsible for creating a log of the requests handled by the server.
- `package.json`: The Node.js project definition. Here, you can find the name and version of the project and any dependencies.
- `payments.json`: The OpenAPI definition that you have created.
- `README.md`: Documentation explaining what the generated code is and how to run it.

- `services`: The services that have been generated, including code where you can implement the business logic for handling payments.
- `utils`: A folder that contains generic utilities that are used by the server.

Now, try running the server and opening `http://localhost:8080/api-docs` on your favorite web browser. You will be surprised to see that the generator created a fully working API portal for you. You can even execute requests from the working portal and see the responses on your browser:

Figure 9.7 – A fully working API portal generated from an OpenAPI definition

As you can see, creating a server using OpenAPI Generator is more complicated than doing so with Postman. However, it offers more possibilities for customization. What matters is that you have a running prototype of your API that you can transform into a full product by adding the missing business logic. Also, you could easily generate the API server code in another language, such as Python or Java, by following the same steps.

Summary

At this point, you know how to go from having a machine-readable API definition document to being able to make requests to your API server prototype. You also understand that the prototype is the first version of your API product and doesn't include any business logic. You understand that the business logic and all the code can be in any programming language and framework, so long as it fits your criteria. To decide which programming language to use, you know how to use a simple method that compares the learning curve, community, job market, and performance to find the best choice for your business requirements. Finally, you know how to move ahead by experimenting with different API code generation tools during the prototyping phase.

You started this chapter with an API design that had been validated and defined using a machine-readable document. With that definition, you learned how to create a working API prototype. You learned about the differences between API mocking and prototyping. You also understood how to choose a programming language and framework according to your business requirements. To do that, you learned how the most popular programming languages rank on factors such as their learning curve, community, job market, and performance. Then, you learned how to generate working API server code from your existing machine-readable definition document. You understood that the generated code doesn't include any business logic, and its goal is to be the first version of your API product.

Here are some of the topics that you learned about in this chapter:

- API mocking is used mostly to validate an API design, while prototyping is used to create the first version of an API

- Mocking uses simulated responses obtained from fake data

- Prototyping doesn't require a full implementation leaving out any business logic

- How you can use off-the-shelf solutions to enhance your API prototype, especially in areas that can be handled by third parties

- How to choose a programming language and framework that you can use to build the API you designed

- The different types of learning curves and how they influence the development of an API product

- Using factors such as learning curve, community, job market, and performance to measure how a programming language adapts to your business requirements

- The most popular programming languages and frameworks for building APIs: Node.js, Python, Ruby, Java, Go, and Rust

- How you can generate server code from an existing machine-readable API definition document

- How you can use Postman to generate an API prototype using Node.js and the Express framework and how you can follow a similar approach using OpenAPI Generator

- How you can quickly have an API portal up and running by using OpenAPI Generator with an OpenAPI definition document

- The details of the generated code, including touch points where you can add business logic

- How easy it would be to generate the same prototype using another combination of available programming languages and frameworks

At this point, you have a working API prototype that you can test yourself and share with your stakeholders. Once you're satisfied with the prototype, you or your team can add the missing business logic. Then, your API is ready for the next step, where you'll make sure it follows all the security best practices. In the next chapter, you'll learn how to implement authentication and authorization in your API as well as understand how to test for common security holes. Turn the page and keep reading to learn more.

10
API Security

Most existing APIs are considered insecure. At least, that's what API security experts agree with. According to Cequence Security, an API security vendor, **account takeover (ATO)** attacks on APIs increased by about 62% in the second half of 2021. ATO is just one of the most common types of attack vectors that can affect your API. These are usually related to cryptographic failures or the lack of secure storage and transmission of sensitive information. Making sure that your API is designed with security in mind is critical to protect you against attackers.

This chapter will begin by defining what API security is. First, you'll get to know how to design secure APIs. You'll then learn that software security is a well-studied area where vulnerabilities are openly shared. You'll learn about the **Open Web Application Security Project (OWASP)** and its list of top API security risks. After that, you'll see the different ways to test your API for security gaps. You'll learn the differences between static and dynamic security testing. You'll also learn how to use security rulesets to test your API and how input validation testing can help you mitigate unanticipated risks. You'll then learn what fuzzing is and how it can be an effective way of doing a security stress test on your API. After that, you'll learn about authentication, including methods such as API keys, OAuth, and JWT. You'll also get to know the details of API key management and secure token generation and storage. You'll then learn about role-based authorization and its ability to perform fine-grained access control. At the end of this chapter, you'll see how OAuth scopes can be an option to implement granular access control for your API.

After reading this chapter, you'll undoubtedly know what API security is and how important it is to the success of your API product. You'll know how to keep updated on the latest security vulnerabilities and how to make sure that you're safe by running periodic testing. You'll also know how authentication and authorization play a significant role in the security of your API. In detail, you'll know how to manage API keys to guarantee that they're generated and stored securely. You'll also know how to define role-based authorization for your API and how to decide what roles to use. Finally, you'll understand how you can use OAuth scopes to let users authorize third-party permissions granularly.

In this chapter, you'll learn about these main topics:

- What is API security?
- Authentication
- Authorization

What is API security?

The ability to design secure APIs is something that involves multiple disciplines. You need to consider several details that can open your API to potential attackers who can access and manipulate sensitive information. The investment pays itself, though, because the cost of addressing a security breach can undermine your API product. The most important areas that you should pay attention to include authentication and authorization, input validation, rate limiting, token and API key management, monitoring, and education. This last item is particularly important because it increases your team's awareness of API security and puts everyone on the same page. Fortunately, there are entities dedicated to security education and awareness.

OWASP is a non-profit organization that works on helping people and companies become more aware of software security. OWASP offers tools and educational resources that you can use to guide your API design and testing. One of the things that OWASP offers is its "Security Top 10 list." They offer a list of the 10 security vulnerabilities they consider the most critical. Their *API Security Top 10* for 2023 is particularly interesting to learn about API security and address the most important vulnerabilities during the design stage. These are the 10 items that OWASP considered the most important in 2023:

- **Broken object-level authorization**: The ability that attackers have to access and manipulate objects without authorization to do so. This can happen if the API isn't verifying if users are authorized to access and manipulate the objects they request.

- **Broken authentication**: Users can bypass authentication by performing multiple attempts, usually employing a brute-force attack. This can happen if the API permits weak authentication or accepts unauthenticated calls from other APIs with weak authentication.

- **Broken object property-level authorization**: API consumers can access and manipulate object properties without authorization to do so. This happens when the API can expose the properties of an object without checking if the consumer has reading permissions. Additionally, consumers can change the value of object properties without writing permissions.

- **Unrestricted resource consumption**: Users can deplete the server's resources, such as memory and bandwidth, by making unrestricted requests. This happens whenever there are no limits, such as regarding the maximum upload file size, timeouts, number of operations to perform on a single request, and number of records to return on a request.

- **Broken function-level authorization**: Regular users can access functions that aren't available to them. This situation occurs when the API doesn't verify if consumers are allowed to perform the functions they request.

- **Unrestricted access to sensitive business flows**: The ability to perform actions through the API that can put the business at risk. This happens whenever the API doesn't check if a combination of requests can lead to a threatening situation.

- **Server-side request forgery**: This occurs when the API accesses a remote resource without first validating it. This can happen when consumers are allowed to store unverified URLs that are then accessed by the API server.

- **Security misconfiguration**: This happens when users exploit known software vulnerabilities to perform unauthorized actions and possibly access and manipulate sensitive data. This can occur when the server software is not up-to-date, the latest security patches are not applied, and unnecessary features are publicly exposed or mismanaged.

- **Improper inventory management**: This occurs when different versions of the API have different authorization and control mechanisms in place. Consumers can bypass authorization or other forms of control by accessing a version of the API where checks aren't in place.

- **Unsafe consumption of APIs**: This happens when the API interacts with third-party APIs directly. If a third party changes its API in unanticipated ways, the data that's shared can create a security risk.

Paying attention to and following the OWASP API Security Top 10 gives you information to fix potential issues. It also gives you knowledge of the things you need to test to make sure your API is sound from a security perspective. Each of the threats has a level of detectability that measures how hard it is for you to test it. While some risks are considered by OWASP as easy to detect, others, such as improper inventory management, have a more complex detectability. Let's see how you can test your API for commonly known vulnerabilities and detect potential security risks.

Security testing

There are three main types of API security testing that you can use to detect vulnerabilities: **static application security testing** (**SAST**), **dynamic application security testing** (**DAST**), and fuzzing. SAST is an analysis technique that scans the machine-readable API definition document and the API's source code to look for known vulnerabilities. On the other hand, DAST performs tests against a running API to detect potential risks. Fuzzing is a type of DAST that uses combinations of input data to test the API for improper input handling and validation. Read on to see in detail how SAST and DAST work.

SAST

While SAST was originally designed to analyze source code, it can also be used to scan anything against a known set of rules – and that includes the machine-readable API definition documents you learned about in *Chapter 8*. You can use SAST to scan OpenAPI documents or any other text-based machine-readable API definitions, so long as you have a predetermined set of rules. Using tools such as Spectral, you can scan your OpenAPI definition to detect any of the OWASP risks you read about earlier in this chapter. However, while analyzing your API definition is a first step, it's not enough. That's because many risk-related patterns can occur in your API code base.

To perform a SAST scan against your code base, you can use tools such as 42Crunch, the AWS Automated Security Helper, or Snyk. Most available tools will produce a comprehensive report that will give you information on how you can mitigate the detected risks. Snyk, for instance, gives you a high-level summary of what was detected and also a detailed analysis of each vulnerability. It also shares detailed information about each vulnerability, such as the complexity of the eventual attack, its criticality level, and its vector. There are thousands of reported and updated vulnerabilities available openly and maintained by the community.

If one of the advantages of using SAST is to detect risks and obtain mitigation information, another of its benefits is to automate its execution. This can be done by configuring your SAST tool to run whenever there's a change in the source code of your API or its definition and just before there's a deployment attempt. By performing a security analysis during your continuous deployment process, you'll have the guarantee that the code that is deployed is free of security vulnerabilities. At least you know that your API code isn't introducing any known risks. However, any code that is running can be prone to new vulnerabilities that were unknown at the time of the static scan. You should also be able to detect and understand those new vulnerabilities, and that's exactly what DAST is for.

DAST

The dynamic nature of DAST lets you perform security testing in a real-world scenario. You can run a security scanner against any of the environments where you have your API server running. You can, for instance, do preemptive security scanning against a staging environment and periodic scanning against your main production environment. DAST tools work by simulating attacks from a list of known security vulnerabilities. They don't have access to your source code or API definition document. Instead, they behave as if they were a regular API consumer interacting with your API.

One of the tasks that you can perform with a DAST is API feature scanning. The scanner tries combinations of possible ways to interact with your API and reports a list of discovered features. Those features can be HTTP endpoints in the case of REST APIs or any other methods that are exposed when you follow a different architectural style. One of the benefits of performing a feature scan is confronting the list of what is exposed with what you defined and seeing if there are any differences. Not knowing what is exposed can also be a security risk.

DAST can also perform several types of testing on your API, looking for patterns that can introduce vulnerabilities. These are some of the types of tests that you can perform with a DAST tool:

- **Input validation**: The scanner attempts different kinds of input against your API features to see how it behaves. It can detect if your API is open to vulnerabilities such as **cross-site scripting (XSS)** and SQL injection.

- **Authentication and authorization**: The tool tries to bypass authentication or use commonly known weak credentials. It also attempts to access and manipulate unauthorized resources to identify escalation vulnerabilities.

- **Session management**: The scanner verifies if your API is handling and managing sessions correctly. It tries to impersonate other API consumers by using different session identifiers and checks if the mechanism used to maintain the session is secure.

- **Injection and manipulation**: The tool tries to use data known to create issues with the XML or JSON formats. The goal is to verify if the API handles the malicious input data properly and correctly defends itself against injection attacks.

- **Information leakage**: The scanner attempts to make requests that generate errors to check if the output leaks any private information. By attempting to send malformed information and access non-existent features, it can understand what type of information is being leaked by your API.

As you can see, after you run these tests, you can gain deep knowledge about your API vulnerabilities. The next step is automating the scanning so that it can run periodically against any environment and alert you whenever a risk is detected. Running the scanner periodically is particularly important because new security vulnerabilities are discovered regularly. To give you an idea, the number of records reported by the **Common Vulnerabilities and Exposures (CVE)** list in 2023 is over 200,000. Every month, over 2,000 new vulnerabilities are detected and reported, making periodic scanning a critical activity.

There's another type of security scanning that is similar to DAST but can produce different results. Keep reading to learn about fuzzing and how you can use it with your API.

Fuzzing

While DAST focuses on well-known vulnerabilities and security risk patterns, fuzzing can test your API behavior while handling unexpected inputs. You can see fuzzing as a type of DAST that focuses on measuring your API's ability to react to unanticipated requests and inputs. With fuzzing, you'll be able to understand if there's any missing input validation or, even worse, if the way your API verifies input data is not correct.

Fuzzing applies algorithms that can manipulate input data to produce unexpected information that puts your API to the test. As an example, most fuzzing tools change input data by adding text, altering the order of characters, or using special international characters. It can also detect issues related to memory consumption by sending input data larger than the maximum limit.

Good fuzzing tools, or **fuzzers**, don't simply produce random data to test input handling methods. They're smarter by detecting the input type and generating the right kind of information. For example, testing a credit card input for a payment API would involve sending random – yet valid – credit card numbers. Additionally, the fuzzer would send other types of data with the same length as the credit card number, as well as numbers exceeding the maximum limit. The report of this test would include information on how the API reacted to each input, highlighting those situations where server errors were triggered or compromising information was shared in error messages.

By applying fuzzing techniques to any API input field, you'll understand if your server implementation is vulnerable to security risks. Of particular importance are features related to authentication and authorization. Authentication verifies the identities of API consumers and provides access to protected features. API consumers have to prove they are who they claim to be by using an authentication method such as an API key. Authorization is a different concept whose goal is to guarantee that only certain authenticated consumers can access protected features and resources. Continue reading to learn more about how API authentication works.

Authentication

The first step to understanding who is using your API is being able to identify consumers. The role of authentication is to establish a valid identity for all the users of your API. Each API consumer becomes an identified entity that you can track individually. With authentication, every request to your API can be mapped to a single consumer. Let's see how you can set up your API to enable authentication. First, let's look at how you can authenticate users:

- **HTTP basic authentication**: On each request, the consumer sends a username and password to the server. The data is encoded using the Base64 method, and unless you're using a secure HTTP server, this method is not considered secure because the information can be intercepted.

- **API key**: Consumers send a key along with the other information on each request. The API key can be sent on an HTTP header, the query string, or a specific field. API keys can have any format and are typically an alphanumeric sequence. This method is simple to implement but lacks strong security mechanisms.

- **OAuth**: Even though it's used mostly for authorization, it's also an excellent way of implementing API authentication. Along with OpenID Connect, you can use OAuth to authenticate users without having to store their credentials.

- **JSON Web Tokens (JWT)**: This is a way to represent that a consumer can access an API. The token can store important information about each API consumer, including a way to verify identity using a signature.

There are other ways you can implement API authentication. However, these are the most popular ones and the ones that are easier to implement. Now, let's look at how your API can manage API keys and use them to verify the identity of the consumers making the requests.

API key management

Managing your API keys is critical to having a product that is secure and guarantees that only authenticated users have access. The first element to consider is the generation of each key. Generating API keys is an area that has matured to a point where there are numerous algorithms that you can use without having to reinvent the wheel. The goal is to generate each key so that it's unique and difficult to replicate. You want to have a different key for each consumer and also make sure that potential attackers can't come up with a key randomly.

One potential API key generation algorithm to consider is the **Universal Unique Identifier (UUID)**. It certainly guarantees uniqueness among the API keys of all your consumers. UUID version 4 can generate a total of 103 billion combinations. The probability of finding a collision, or a duplicate, is approximately one in a billion. This probability is considered very small and can thus be neglected.

Another area to consider is the storage of each API key. Because API keys are used as a means to authenticate consumers, you should pay special care when storing them. API consumers shouldn't store API keys in source code or a code repository as they can easily be retrieved by malicious players. You, as the API owner, have the biggest responsibility because you'll have to store the keys of all consumers. To do that, you can use solutions that provide database encryption or secure key vaults. Examples of such solutions include AWS Secrets Manager and Hashicorp's Vault.

Both of those solutions offer a way to revoke API keys. Revocation is an important part of API key management as it disables a key and prevents a consumer using that key from authenticating and accessing your API. You should establish a process for simple API key revocation so that you can easily disable a key whenever you detect suspicious activity or when you want to disable access to the corresponding consumer. An automated way to create an effect similar to revocation is to set an expiration time for each API key. Whenever the expiration time is reached, the key is automatically disabled. This will prevent consumers from using the same key over long periods but requires them to obtain a refreshed key, making the process taxing to your users.

Whatever implementation you decide to use, make sure that you document it thoroughly. API consumers need to know the format you'll be using for your keys, the expiration policy you choose to follow, and the rules you have in place for deciding when to revoke a key. They also need to know how they should present the API key so that your server can obtain it and attempt to validate it. Your documentation should present examples of API requests where the API key is present so that consumers know how they can authenticate. If all these things sound complicated to you, many API gateway products already offer them. Examples of such solutions include Kong and AWS API Gateway. Most solutions that offer API key management can also be used to manage authentication tokens. Keep reading to learn more about token management.

Token management

While managing tokens is similar to the way you can manage API keys, there are important differences to consider. The first one is in the way tokens are encrypted. Because JSON-based tokens can hold sensitive information, you must encrypt their content. Unlike API keys, which represent identifiers, tokens can have any information embedded. Here's an example of an unencoded JWT payload so that you can better understand what type of information it can hold:

```
{
    "amount": 10,
    "currency": "EUR",
    "orderId": "3545c502-c490-4591-8d52-c8592872523a",
    "paymentMethod": "creditCard"
}
```

A JWT is comprised of a header that defines the token type and the algorithm used for its signature. The header is then followed by a payload, such as the one you've seen just now, and, finally, a signature. The signature is used to verify the validity of the token, and this is why it's important to pick a strong secret to encrypt the information. Encrypting the payload is an extra optional step that can increase the security of the data within. While not mandatory, it's something you can consider if the nature of the data inside the token is private or sensitive.

As you can see, it's not easy to tamper with a JWT because it requires a signature that uses a secret. While the data inside the payload can be changed, the signature will determine the validity of the information. A JWT can be used to carry all types of information, including authorization details. Read on to learn what API authorization can control.

Authorization

With authentication, you make sure API consumers are correctly identified, and their access is controlled. Authorization happens right after, and its goal is to establish what authenticated users are allowed to do when accessing your API.

RBAC

One popular authorization model is **role-based access control** (**RBAC**). It works by first establishing a set of roles and then associating roles with permitted actions. Examples of common roles include the "administrator" and the "regular user." Each feature then has to verify what role the API consumer has and if the requested action is listed as permitted for that role.

It's important to highlight that, to be considered effective, RBAC has to be enforced at the interface level and then on each feature that the API server implements. Otherwise, you might end up letting users perform actions for which they don't have the right permission. It's possible to implement RBAC at the interface level by configuring the API gateway to follow an RBAC strategy. You can map roles to API operations and check what consumers can access before they even interact with the server. The Kong gateway, for instance, has RBAC plugins that you can install and use easily. On the other hand, RBAC implementation on the API server depends on your choice of programming language and framework. On each request, you need information about the consumer's identity and their role. Then, you can check if the role has permission to access the requested operation.

The granularity of roles determines how much you can control the permissions on an RBAC system. If you have too few roles, you won't be able to be precise about the permissions you grant to each consumer. If, on the other hand, you have too many roles, your system can become too complex and difficult to implement and maintain. Reaching the right balance between the number of roles and features is critical. One way to look at it is to think about RBAC as fine-grained access control. Here's a recipe for configuring the system, starting with the API resources and ending with the permission evaluation:

1. **Resource mapping**: At the interface level, you identify the resources that need access control enforcement.

2. **Operation mapping**: You list all the operations consumers can perform on each mapped resource. Examples of operations include reading a resource, creating resources of the same type, and removing a resource.

3. **Role definition**: Create roles that encapsulate operations on resources. Each role has a list of operations that it can perform. You can map operations at a high level or the level of each resource. For example, the "administrator" role could have access to removing any resource, while a more restrictive role could have access to removing only a particular type of resource.

4. **Permission evaluation**: On each request to the API, evaluate what role the consumer has and if the requested operation is permitted.

With this list, you'll be able to implement your own RBAC system and make sure that every consumer can only perform authorized operations. An alternative to use at the interface level is OAuth and its scope-based authorization approach.

OAuth scopes

OAuth scopes let you define the permissions that an API consumer requests during the authorization process. OAuth allows API consumers to obtain authorization to act on behalf of users, and during that process, they are given a set of permissions defined by scopes. Users actively need to grant API consumers the permissions defined by the scopes. A good example is the authorization pop-up window you see when you're using a popular service such as Google to sign into a third party. The authorization window asks you to grant the third-party-specific permissions defined by OAuth scopes.

Whenever the third-party consumer accesses your API, the system verifies if the set of granted scopes intersects the scopes required to perform the operation. Like with RBAC, OAuth scopes can have a varying degree of granularity that dictates how easy it is to manage them. OAuth scopes can also be mandatory or optional, making it easy for users to select what permissions they grant to third parties. Overall, OAuth scopes are a good way to provide API authorization by letting users manage the permissions they give to consumers acting on their behalf.

Summary

At this point, you have the high-level information required to understand and communicate API security. You know that a secure API begins with a thoughtful design. You also know that thinking about API security requires a holistic view of how consumers interact with your implementation. API security is as much about authentication as it is about testing and being updated about the current vulnerabilities. You know about OWASP and how to find reported security vulnerabilities as they're discovered. You also know that a fine-grained authorization system is an excellent measure to mitigate security risks. Finally, you know how to combine the different security components to offer a successful API product.

This chapter began by identifying the key areas of API security. You learned about the OWASP "Security Top 10" list and how to be on top of newly discovered vulnerabilities. You then learned about API security testing methodologies, including SAST, DAST, and fuzzing. Then, you learned about API authentication, its most important variants, and how to manage API keys. You also learned how to generate API keys to avoid potential collisions and prevent brute-force attacks. Finally, you learned about authorization using RBAC and how roles and permissions work. In summary, this chapter taught you how to manage good API security.

Here are some of the topics we covered during this chapter:

- API security, which is the process of designing, implementing, and maintaining APIs that don't pose security risks

- The most critical areas of API security, which are authentication and authorization, input validation, rate limiting, token and API key management, monitoring, and education

- How to use the OWASP " Security Top 10" list to keep track of potential vulnerabilities

- The three main types of API security testing: SAST, DAST, and fuzzing

- Using specific rulesets, you can use SAST to test your API for security risks

- The importance of running periodic API security tests as the list of known vulnerabilities is constantly changing

- How DAST can help you find security risk patterns, even in parts of the API that seemed safe

- The types of tests that a fuzzing tool performs

- The most important types of API authentication: HTTP basic authentication, API key, OAuth, and JWT

- How to generate and store API keys securely

- The importance of encrypting information shared using JWTs

- How to configure the permissions of users using RBAC

- The influence of the granularity of roles on the maintainability of the authorization system

- How OAuth scopes work

At this point, you know how to apply measures to your API product to ensure it has no known vulnerabilities. Those measures include authentication, authorization, and periodic security testing. You're now ready for the next step in testing, which includes not only security but all aspects of your API. Keep reading to learn all about API contracts, performance, and acceptance testing.

11
API Testing

Building an API product without testing it is like trying to cross the ocean without a compass. Testing is a crucial activity that helps you identify issues related to performance, security, functionality, and also alignment with your stakeholders. API testing is a way to guarantee that you're building what you've planned. It's also a way to confirm that what's running is what you have defined and that it works as expected.

You begin this chapter by exploring contract testing. You'll learn that the goal of contract testing is to make sure that your API is behaving consistently. You'll then learn how to define a contract test and find out what API elements can go into the definition. You'll learn that you can not only test requests and responses but also the structure of payloads, data serialization formats, and headers, among other things. After that, you'll see how having a well-written machine-readable definition is fundamental to running contract tests. You'll learn how to use Pact, a tool that helps you capture and store API traffic that can later be used as the source of repeated contract tests. In contrast, you'll also see how to use Postman, another API tool, to configure and run contract tests. After that, you'll jump into performance testing and learn how it can help you plan your infrastructure even before you launch your API. You'll understand how you can test and measure the load, response times, and resource usage of your API infrastructure. You'll take a deep look at Apache JMeter, a tool dedicated to web performance testing, to find out how scalable your API is. Finally, you'll see what acceptance testing is and how you can combine security, contract, and performance testing to obtain the results you need. Among other things, you'll see how you can translate a job-to-be-done into a functional acceptance test.

At the end of this chapter, you'll know the different kinds of API testing. Not only will you know about security testing, which we covered in the previous chapter, but you'll also know about API contract, performance, and acceptance testing. You'll know how to define, run, and interpret a contract test. You'll also know how to measure the scalability of your API by applying different request loads while performing a performance test. In the end, you'll have the knowledge that acceptance tests are a combination of other tests to guarantee that the API behaves according to your business requirements. Finally, you'll know how tools such as Postman, Pact, and Apache JMeter can help you define, run, and analyze your different API tests.

This is divided into the following main topics in this chapter:

- Contract testing
- Performance testing
- Acceptance testing

Contract testing

APIs are the glue that allows the integration of different applications to create a rich experience. Maintaining the reliability and consistency of integrations is crucial to the success of an API product. You can think of your API's machine-readable definition as a contract that dictates how your API behaves. You can also use that contract to check whether the behavior changes. This is what's called contract testing. Keep reading to learn all about it and how to put it to work.

Essentially, you can define API contract testing as a set of scripts that, when executed, test how well an API conforms to what's written in its machine-readable definition. By performing contract tests regularly, you ensure your API's behavior is precisely as it's documented and expected. Knowing that the API behaves as expected increases the confidence of consumers and decreases the chances of issues occurring during an integration. Let's look at the elements that you can define on an API contract that can then be tested for consistency:

- **Requests and responses**: The structure of API requests, HTTP requests and response headers, query parameters, body content, HTTP methods, response structure, and HTTP status codes.

- **Payload structure**: The format of the payload exchanged between the API server and its consumers. It includes the payload data fields and types, information on what optional fields are there, and any existing objects and arrays.

- **Error handling**: What happens when an error occurs. Specifically, the error message that is shared with the API consumer, its structure, the information therein, and the HTTP error code used in the message.

- **Authentication and authorization**: The types of authentication that the API implements, the authorization roles, and their permissions.

- **Rate limiting**: The way rate limits are applied to consumers with different roles and what happens when limits are exceeded.

- **Data validation**: Any constraints on input data such as field lengths, allowed input formats, and any validations that happen once data is sent to the API server.

- **Hypermedia and linking**: The structure of hypermedia in REST API responses and how consumers can navigate between resources using the provided links.

- **Webhooks**: The definitions of webhook URLs, any payload structures for asynchronous notifications, and the shape of the information sent from the API to the webhook.

- **Versioning**: Rules for maintaining compatibility with previous versions of the API and how different versions are described and communicated.

- **Response time and performance**: The acceptable time it takes the API to respond to different requests and the performance of the API under various loads.

- **Data serialization**: How the API uses different serialization formats, such as JSON and XML, and the rules for formatting data in requests and responses.

- **Headers and content negotiation**: The support request and response headers and the way the API handles content type negotiation.

Now that you know what you can verify during a contract test, let's see how you can implement one. The very first thing is to make sure that you have a suitable contract that you can use to test the elements that you care about. If you're using OpenAPI, you already have half of the work done for you. You just need to confirm that the OpenAPI machine-readable document has all the definitions you want to test. For example, if you're going to test content negotiation, you need to have it defined in your OpenAPI document.

The second thing you want to do is automate your contract testing. By automating the execution of contract tests, you ensure that anyone can test your API with a minimum of effort. Automation also has the advantage of helping you guarantee that your API is consistent and follows the contract at all times. One way to make sure that contract tests run every time something changes is to integrate the tests into your deployment pipeline. By making the contract tests run every time someone attempts to deploy a change to production—or another important environment—you can make the deployment depend on the result of the test. If the contract test fails, then the deployment won't happen. You'll then have the opportunity to fix the API and make sure that the changes you're introducing aren't breaking the contract.

However, there are situations where you actually want to introduce changes to your API, and those changes will break the contract. Even worse, sometimes, those changes will create inconsistencies and break compatibility with previous versions of the API. To accommodate those situations, you have to change the contract tests so the API doesn't fail with the introduced changes. How to manage and communicate the changes to your consumers is a different topic that you'll read about in detail in a coming chapter. The summary is that you should deal with any change from the perspective of API design and engage with all stakeholders while introducing it.

Working with API contract testing is only as effective as the tools you use. While you can perform manual contract testing with almost any tool that lets you perform an API request, implementing full automation requires a prescriptive approach. Pact is an example of a tool that enables you to configure, execute, and automate your contract tests. Beth Skurrie originally created this tool as a way to solve the challenges related to testing API interactions. It was first released as a Ruby gem in 2013 and evolved into a cross-language framework available in popular programming languages such as Java, JavaScript, and Python. Pact has had an increase in adoption since its inception and is now backed commercially by a business.

Pact works by capturing API requests and storing them using a mock server. The captured requests are stored and can be replayed as many times as you want against the real API. The goal is that the requests that constitute the contract test are as close to reality as possible without creating a dependency on a real consumer. This approach makes setting up and running a contract test very straightforward and simple to maintain.

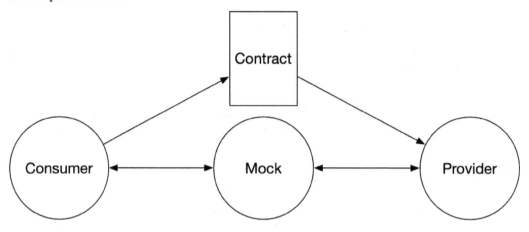

Figure 11.1 – A simple flow diagram of a Pact test

A different approach is the one that Postman follows. The popular API collaboration tool can translate a machine-readable OpenAPI definition into a series of requests that perform a series of live tests against your API. The goal of the tests is to verify how much the real API conforms to what's defined in the machine-readable document. Because the tests are stored in a Postman collection, they can be easily reused and executed by anyone in the organization. Speaking of Postman, you can also use it to run API performance tests. Keep reading to learn all about testing your API for performance.

Performance testing

If contract testing allows you to ensure that your API behavior is consistent over time, performance testing gives you the information you need while tuning your API to deliver the best possible consumer experience. In *Chapter 2*, you learned that reliability and performance are fundamental to a positive API user experience. In particular, having an API that responds well under growing load situations is critical. As you're seeking success for your API product, you want the number of consumers to grow, and your API has to stay performant.

Actually, both scalability—the ability to stay performant even with increasing levels of usage—and responsiveness are the two things that you verify during a performance test. Scalability can be tested by increasing the usage load in a controlled fashion and seeing how the response behavior changes. In each step of increasing load, you test the API responsiveness and measure how well it can process the same requests and how it responds to the consumer. The goal is for the API's responsiveness to remain unaltered through varying degrees of usage, which, by definition, makes it scalable. Let's see what you can test to verify the API's scalability:

- **Load**: The number of concurrent requests to the API. Simulate an increase in load and discover how the API behaves under heavy usage.

- **Response time**: The time it takes for the API to respond to a single request. Reasonable response times vary depending on the API features. Measure the response time under different loads to understand how your API behaves and adapt it to meet the needs of consumers.

- **Stress**: How the API behaves beyond reasonable usage load conditions. Simulate a dramatic increase in load and see when the API crashes or when its response time increases significantly.

- **Resource usage**: How resources such as CPU and memory are used whenever the API is accessed. Simulate an increase in load and measure the resources used by the API.

As you can see, everything starts with manipulating the load to simulate different operating conditions. After that, you can measure different things and use the results to improve how the API works. In fact, interpreting the results is often as important as running the tests. Let's now look at how you can implement API performance testing by understanding what tools are available and what kinds of test scenarios you should be trying to simulate.

As mentioned, while exploring contract testing, one of the tools that you can also use for running performance tests is Postman. It works by letting you run the requests from a collection repeatedly. You can simulate the number of concurrent requests by setting the number of virtual users of the API. You can also set up the total duration of the test and how the number of virtual users will load—all at once from the beginning or ramping up over time until the maximum is reached. One thing to keep in mind while using Postman or any other tool is the limits imposed by the tool and the machine where it's running. Keep in mind that the total number of simultaneous requests and the total time of the test are limited by the available CPU and memory of the host running the test.

Another tool you can use is Apache JMeter. Unlike Postman, which has many features, JMeter is dedicated to performance testing. It's been around since 1998, when it was created inside the Apache Software Foundation by Stefano Mazzocchi. Its popularity has been increasing over time, and in 2023, it was used to run all kinds of performance tests on static and dynamic web resources and applications, including APIs. It's an open source tool that you can use freely with an engaged community and plenty of documentation available online. To use it, you first create a test plan that you can then use to execute the performance test. The easiest way to create a test plan is to launch the JMeter application on your desktop and then set up how you want your performance test to run.

One exciting thing about JMeter is that you can define all the options for your performance test in great detail. You can, for example, determine the number of concurrent API consumers and the shape of the curve used to ramp up the parallel requests. You can test APIs of different architectural types, including REST and SOAP. You can configure API authentication and even have different concurrent consumers authenticating with different credentials to better simulate a real scenario. Finding scenarios that are as close to reality as possible is critical to how your tests will be able to help you improve your API.

Designing a realistic API performance test scenario starts with understanding how potential consumers will interact with the API. Your goal is to identify and document use cases and the requests consumers make to fulfill their needs. Then, create a test plan that includes all the requests for each use case. Those requests should be completed sequentially and in the correct order to mimic consumer behavior. An even better approximation of reality includes adding random delays between requests to simulate network and consumer processing latency. After you have a reasonable number of use cases, you can set up your testing tool to execute the test plan for each use case concurrently. This will simulate one consumer in each use case. Finally, calculate how many consumers you want to test and set up a distribution between use cases so they all run in parallel. Something to keep in mind when executing tests that can generate a heavy load is to do them against an environment that you can control. Keep reading to learn how to set up test environments.

The worst thing that can happen during a performance test is having your production environment knocked out under a heavy load coming from the test requests. To prevent that, it's critical that you have a dedicated test environment and a separate test scenario that is more aggressive than the one you run on a stage or production environment. Building an environment that looks like your production can be challenging. First, you need to catalog all the services that are running in your production environment, not just the server where your API code is running. Any other services, such as databases, also need to be replicated in the test environment. The second part is replicating—or generating—the data that you'll load into your test environment. If the goal is to have an environment that's similar to production, your data must also be similar. That means you'll need to load the test environment with a reasonable amount of user data, including credentials and data that consumers will use during their requests.

All those requests, along with the time it took your API to respond and other related metrics, become available as a report of each performance test. Interpreting the results becomes your priority to understand if there's something you need to improve on your API environment. So, for example, to understand how your API would perform under a peak load, you'd look at the results of a test that mimics that scenario. If the response time remains consistent across different growth in usage, then your API is handling the load just fine. If, on the other hand, the response time or the number of errors in the responses grows, it means that you need to improve your API's performance. You can determine the load that generates the bottleneck on your API server and use the information to make the necessary improvements. The final type of testing that you'll see makes sure that the API meets all the business and technical requirements. Read on to learn about API acceptance testing.

Acceptance testing

Knowing that your API has consistent behavior from an integration perspective and being sure that it can scale to peak loads are two crucial aspects of building a successful API product. A third aspect is knowing that the API is consistently behaving according to your business and technical requirements. And that's where acceptance tests come in. With acceptance tests, you can map the expectations of all stakeholders to the behavior of the API through a set of scenarios that can be repeatedly verified. You can not only test the API for situations where the results are positive but also identify situations where you know the responses contain errors and verify them through test cases. Overall, acceptance tests can guarantee that the API design that you've made is being applied and the API behaves according to it consistently. Read on to learn how to make the best of acceptance testing for your API product.

Similar to other types of tests, defining a comprehensive set of cases is crucial to obtaining meaningful results from acceptance tests. You should return to the user research you did while designing the API and use the **jobs-to-be-done (JTBDs)** you identified as a base for the acceptance tests. The goal is to test how well the API implements the features that fulfill those JTBDs and how consumers see it as the solution to their challenges. This is what is called functional acceptance testing. The tests verify whether the API functionalities are the ones that consumers are expecting. Let's look at an example to see how it works.

A simple JTBD you analyzed in *Chapter 5* was "paying for an online service quickly and easily." Back then, you were identifying potential users of a web payments API. Knowledge of the JTBD resulted in the definition of use cases in *Chapter 7*, along with the flow that consumers would perform while accomplishing their jobs. For the example we're using, the flow consisted of several steps involving a few requests to an API sharing information back and forth with the user's web browser. The acceptance test for this example API can derive from the flow to make sure that the requests work and that both the request and response payloads allow consumers to accomplish their JTBD. Moreover, the acceptance tests should also verify the format and shape of the payloads according to the API definition you went through in *Chapter 8*.

While testing for positive behaviors and responses is fundamental, adding tests for negative situations augments the guarantee that the API is acting as you defined it. Negative tests verify that whenever you send a request that is not expected, the API responds with the appropriate error and doesn't crash. For example, in a payments API, if you send an invalid credit card number in a request, you should receive an error in the response. The negative acceptance testing ensures that the API isn't passing any erroneous information into the business logic. Instead, it's short-circuiting the problem at the interface level, as it should.

There's a second kind of acceptance testing where you're not verifying the functionality of your API. In non-functional acceptance testing, what you're doing is checking whether things such as performance, security, and reliability are aligned with your business requirements. Instead of testing each behavior individually, you can use a combination of the non-functional tests you've been learning about. So, for example, to test security-related aspects, you can use techniques such as SAST, DAST, and fuzzing. Likewise, performance can be tested using any of the methods mentioned earlier in this chapter.

To execute acceptance tests on your API, you can use a dedicated environment if you're dealing with tests that don't create a negative impact. If your tests aren't making any changes to the data or introducing an unreasonable load, you're free to use your production environment. While there aren't any tools fully dedicated to API acceptance testing, you can use a combination of contract and performance testing tools to obtain the results you're looking for. So, if your business requirements include a certain level of performance, you'll want to use a tool such as JMeter. Conversely, if you want to make sure that the behavior of your API is fully aligned with the needs of your consumers, you can use a tool such as Pact. Ultimately, you'll want to capture the results of different tools and interpret them as a whole to measure against your goals.

Summary

You have learned how to set up and use contract, performance, and acceptance tests. You know the only way to achieve consistency and alignment with your stakeholders is to periodically and automatically test your API. You also know what kinds of tools you can use to execute different types of tests. On top of that, you know the importance of having a well-defined API with a machine-readable document ready to be used. You know that acceptance testing is a combination of different types of tests. Finally, you know that being able to analyze and take action based on findings from tests is crucial to the success of your API.

You began this chapter by establishing what API contract testing is and how you can set it up and execute it. You then went on to learn about performance testing. You learned how to gradually increase the request load of your API to measure its scalability and responsiveness. You also learned about tools that you can use to set up contract and performance tests. Finally, you learned about acceptance tests and their importance in aligning your API behaviors with your stakeholders. You learned that acceptance tests are a combination of security, contract, and performance tests.

These are some of the topics that you learned about in this chapter:

- The consistency of an API is crucial to its consumers being able to integrate with it

- An API's machine-readable definition can be seen as a contract

- Contract testing verifies whether an API conforms to its machine-readable definition

- You apply contract testing to API requests, responses, payload structure, error handling, and headers, among other things

- Automating contract tests ensures the consistency of the API over time

- Performance testing gives you the information you need to tune your API to offer the best consumer experience

- By increasing the load of requests in a controlled fashion, you can test your API for scalability and responsiveness

- Having an environment dedicated to testing is fundamental to preventing your production API from crashing

- You can use Postman, Pact, and Apache JMeter, among other tools, to implement contract and performance tests

- Acceptance testing is a combination of security, contract, and performance testing

- With acceptance tests, you can guarantee that your API product is aligned with your stakeholders

- You can translate a JTBD into a functional acceptance test that checks whether a functionality of your API is working as expected

- Testing for negative behaviors is as important as testing for positive ones

At this point, you know the different types of API tests, how to set them up, how to run them, and how to analyze the results to implement improvements. Testing plays a significant role in ensuring that the API is aligned with business requirements and the needs of your stakeholders. It's also one of the pillars of offering a quality API product. And that's precisely what you'll see in the next chapter. So, keep reading to learn about API quality assurance.

12
API Quality Assurance

Quality assurance, also known as **QA**, is the systematic process of ensuring your API product meets quality standards, business requirements, and the needs of consumers. To understand how API QA works, let's identify what quality means. In the world of product management, there isn't a single definition of the meaning of quality. However, it's generally accepted that quality measures a product's usability, how much it serves its purpose, and how it follows existing industry standards. Among other criteria, quality measures how a product meets performance, functionality, and reliability.

This chapter starts by defining in detail what QA is and how it's related to standards, business requirements, and the needs of consumers. You'll start by exploring the meaning of quality. Then, you'll learn about the attributes of a high-quality product. After that, you'll see how QA contributes to building a high-quality API product. You'll then learn how running a QA process helps you meet the needs of API consumers. After understanding how QA can help you build an API product, you'll learn how to plan the tests you'll run during the process. You'll learn that the first thing you need to do is define the goals you have for your QA process. After that, you'll learn how those goals help you identify the scope of your tests. You'll then see the relationship between test scopes, environments, and purposes. After that, you'll learn what behavioral testing is and why it plays an important role in API QA. You'll see how you can combine different types of tests to perform behavior analysis. Among other things, you'll learn what boundary value analysis is, and the importance of stateful tests. After that, you'll learn about regression testing and how it helps you make sure that the changes you introduce to your API over time don't generate defects. You'll also learn how to mitigate defects that are discovered during a regression test. Finally, you'll learn about API monitoring and how to set it up. You'll discover different types of monitors that you can create for different purposes. In particular, you'll learn about uptime, performance, security, and usage monitors. To end this chapter, you'll learn that you can convert any type of test into a monitor.

By the time you finish reading the chapter, you'll know what the elements of API QA are and have enhanced your knowledge of testing that you already had from previous chapters by learning how to combine several types of tests to obtain actionable metrics. You'll have gained knowledge about the environments where you can run different types of tests and what their purposes are. You'll also get to know how behavioral and regression tests can improve the quality of your API. Finally, you'll know what API monitors are and how you can take advantage of them to guarantee a stable high-quality API product.

This chapter covers the following topics:

- Definition of QA
- Behavioral testing
- Regression testing
- API monitoring

Defining QA

Peter Drucker, a renowned management author, defined quality in a way that I identify with. To Drucker, the quality of a product is the sum of all its qualities. But quality is also a way to identify how well a product is accepted by its consumers. A high-quality product is one where there is little or no friction. Users of the product can apply it to the problems they want to fix without having to adapt themselves. The better users engage with your API product, the higher the chances of success will be. That's one of the reasons why QA is crucial to achieving your business objectives. Let's look at the three main criteria that QA measures.

To begin with, QA measures a set of standards that the industry where you're operating considers fundamental. So, for example, if you're building a payments-related API, you're operating in the payments service industry. In the European Union, one of the payments industry standards that your API should adhere to is PSD2. QA, in this example, measures how well your API follows the PSD2 standard. The QA process involves testing the API for functionality and behaviors related to the standard that you want to study. You can measure the adherence to industry standards by using API contract testing since it focuses on adherence to a definition – in this case, the industry standard.

Another area that QA verifies is related to business requirements. The goal of the QA process is to ensure that all the features, interactions, and behaviors of the API align with any existing requirements and user stories. Here, you can not only test the API itself but also how other parts of your business react to interactions from consumers. For example, on a payments API, you would want to test how the payment gateway receives requests and if those requests are valid. To execute those verifications, you can use acceptance testing, as it can test both the functional and non-functional attributes that extend beyond the interface into other parts of the system.

Finally, QA verifies if the needs of consumers are met. The QA process tests if the API delivers a satisfactory user experience by checking usability, security, responsiveness, and other non-functional attributes. Here, the focus is on things that are not as tangible as the functional elements of the API but contribute largely to the way consumers perceive and interact with the product.

One of the outcomes of having a QA process is being able to identify defects proactively in an organized way. The worst thing you want as the owner of an API product is to have consumers complaining that things aren't working as expected. By identifying defects before consumers do, you're able to fix them before your API is deployed to a production environment. Fixing defects before they hit users also increases the quality of your API product. Going back to the definition of quality I presented at the beginning of this chapter, it's easy to understand that an API with no defects decreases the friction that consumers have while using it. And, by reducing the friction of the product, you'll increase its acceptance. Now, let's look at how you can plan and execute QA tests.

Test planning and execution

The first thing to think about when planning a QA test is to identify the objective of the test. Being able to understand the goals you want to achieve is critical to having a successful QA process. One way to document your goals is to identify the outcome of the tests. Are you doing QA to make sure all functionality is working as expected, to ensure that your API is fully secure, to optimize performance and reliability, or something else? The objective influences the outcomes, which, in turn, influences the scope of what you'll be testing.

The QA test scope identifies which parts of the API product you want to test. Your role is to define what areas, features, and attributes of the API you want to test. For a REST API, you can identify which paths you want to test and, inside those paths, which HTTP methods. For example, you can define that you'll test all the operations that retrieve data from the API. Or, you can perform QA on all the operations that manipulate information, leaving all other operations untested. What's important is to align the QA test scope with the outcomes you want to achieve. If, for instance, you're looking for a way to increase your API security, then your QA test scope should include security-related items.

Depending on the scope you defined for your QA process, you'll want to run the tests on different environments. The following table provides a summary of different possible API QA test scopes, their preferred environment, and their purpose:

Scope	Environment	Purpose
Operation	Isolated mock server	To verify the correctness of individual API operations
End-to-end	Dedicated QA	To simulate how real consumers would interact with the API
Functional	Tailored to the test scenario	To validate that the tested functionality works according to its specifications
Security	Secure with controlled access	To identify security vulnerabilities
Performance	Dedicated, similar to production	To test the API's performance, scalability, and responsiveness
Regression	Dedicated QA	To check if newly introduced changes produce defects

Along with the different scopes, you can also identify and define different kinds of QA test data. You can use fake data with an API mock server to perform an operational QA or very specific tests on certain parts of the API. However, for end-to-end QA tests, using data that you can reproduce is often a good idea. Being able to run the same QA test with the same – or very similar – data over and over helps you determine what is failing and why with precision. That means that you should have an easy way of creating an environment and filling it with the same data before running the QA test. This process is often called seeding because it establishes the base data that an environment has when it's created.

Now, let's look at two types of QA tests that are often used with APIs.

Behavioral testing

API QA behavioral testing is an approach that has the goal of verifying how an API behaves under different conditions and in different scenarios. Behavioral testing helps ensure that the API works according to its specifications and the standards and regulations it needs to comply with. Behavioral testing is one or a combination of different types of tests, such as the ones you've been reading about. Let's look closely at some types of tests that you were not covered before and how they influence the QA process.

While most testing verifies if the API successfully performs its operations, error handling tests do just the opposite. With error handling tests, you can verify how the API manages errors and exceptions. By intentionally triggering faulty requests, you can see what errors are produced and how the API handles the situation. Your goal is to validate not just the error that is returned for each faulty request – if an error is returned at all – but also to monitor what is happening on the API server. By checking if the API server isn't crashing or behaving erratically, you'll be improving its quality. One way to generate a request that you know will produce errors is to provide an input that you know beforehand not to be valid.

Another type of test uses input values that are valid but are considered rare cases. You make requests that contain input data that will be near the limit of validity. This is called boundary testing and its goal is to ensure the reliability and robustness of your API. Boundary tests help you examine the boundaries or extreme cases of input information and what the API's behavior is in those situations. There are several techniques you can use with boundary tests:

- **Boundary value analysis**: Usually used for numeric input values, this helps you identify off-by-one and rounding errors.

- **Minimum and maximum input values**: This lets you test how your API handles the minimum and maximum allowed values for the data type of each input field.

- **Empty or null values**: By using empty or null values as the input of a request, you can test how the API server reacts. You can apply this technique to numeric and text data types but also to objects and lists.

- **String length**: Each text input – represented by a string – has a maximum length that can be identified on the API machine-readable definition and should be checked by the API server. By sending text that is longer than the specified string length, you're able to verify how the API handles the situation.

- **Date and time**: This is an important area if your API handles date and time input values. You can test minimum and maximum date and time values, values that seem syntactically correct but don't make sense, and also values that aren't syntactically correct. On top of these cases, you can also test how the API handles leap years, differences in time zones, and daylight savings.

Notice how none of the previous techniques require the notion of state to exist between requests. That's because they test isolated operations, as opposed to testing a sequence of actions that are all part of a use case. To be able to test how an API handles information being shared between requests, you need to perform what we call stateful tests. The first step to working with stateful tests is to identify the API's different states and map those states visually to use cases. Let's use our web payments API as an example and map the states related to the payment operation. Whenever a user attempts to make a payment using the API, several steps need to happen. First, the user needs to input their credit card details. Then, the system verifies if the credit card is valid and if it has enough credit to pay the amount due. Finally, the system contacts a payment gateway to execute the payment. In the end, the payment can be successful or denied, depending on the credit card and the amount involved. As you can see, different states are involved in between these steps. Let's see what this use case looks like with a state diagram:

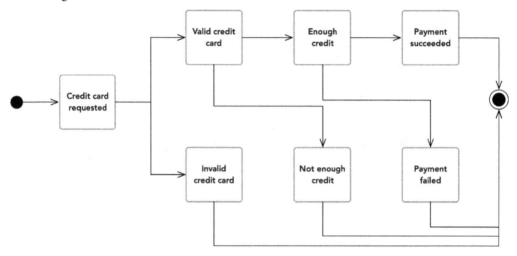

Figure 12.1 – A simplified state diagram of a payment

As you can see, identifying and understanding the different states – and the transitions between states – an API can be in is critical to knowing if the API is working as expected. To test state transitions, you can use functional testing techniques, where you verify if each step of the whole use case functions as expected. You'd also want to run negative tests to check if the API falls into an error state if the request is known to produce a failure. To test the whole use case – and its different state variants – you must run an end-to-end test that combines the different transitions.

Making sure that use cases are handled correctly in positive and negative situations is important to guarantee the quality of your API. Something even more important is confirming that all the API's functionality keeps running over time, even when you introduce new features or fix any defects. Keep reading to learn how you can do that.

Regression testing

So far, you've been seeing ways of testing if what you're building is behaving according to expectations. What you'll see now is a way of checking if any of the previously tested items stopped behaving in the way they were before. Something going from a state of success to a state of failure is known as regression. This type of testing is called regression testing because it detects those situations when something that had been fixed stopped working.

The first thing to pay attention to is the scope of the regression tests. Here, you can limit your tests to anything that has been fixed or you can attempt to test all the existing features. While the former option is easier to set up, maintain, and scale, the latter is more comprehensive and gives you more confidence about the quality of your API. Whatever option you choose, make sure it provides you with the information you're looking for. For example, if your priority is to make sure that you're not breaking existing features whenever you launch something new, then you'll want to give more focus to a comprehensive test scope. On the other hand, if you're interested in guaranteeing that fixed bugs aren't reappearing, then you'll probably be fine with a smaller scope that includes only the existing fixes.

One key area of regression testing is automating the process. By adding automation, you'll not only eliminate the time it takes to run the tests but you'll also put yourself in a situation where you won't ever forget to run them. To be effective, you should configure the automation to run whenever any changes are introduced to the API and just before those changes get deployed to production. That will put the regression tests right at the end of the API development pipeline, making sure that the team responsible for the changes will get visibility into the test report. If all regression tests pass, it means that the changes didn't introduce any known defects. Otherwise, something should be fixed or removed from the production release. In that situation, you have a few options:

- The team responsible for the breaking changes fixes the issue and a deployment to production is possible when all tests pass. This is usually called a hotfix and is done for defects that are considered small and easy to fix.

- The team behind the defect decides to remove the changes from the release and work on a fix. This action is taken whenever the amount of changes is too large or the fix is complex and might take a long time to finish.

- If the defect is on a feature that a different team owns, the team that introduced it removes the changes from the release. After that, both teams work together to find an alternative approach to the introduced changes so that regression doesn't occur.

The types of tests that you can use during regression analysis are all those that you've been reading about in this book. Functional testing, however, has a special role in regression analysis because its goal is to verify if individual features work as expected.

Nonetheless, if the changes you're introducing are related to other areas, you should use whatever type of test is the most appropriate. For example, if you've introduced changes that are related to performance, you should execute performance tests to confirm that the API hasn't lost performance due to the changes. Similarly, if what you're changing has to do with security, you can perform security tests to confirm that all related functions are still working as intended. Authentication, for instance, is a security-related feature that often goes through changes and improvements.

To summarize, regression tests offer many benefits that you wouldn't get otherwise. From a quality perspective, regression tests are one of the best tools you have to guarantee that your API doesn't have any defects. When it comes to regression testing, business stakeholders will find a way to minimize the risks associated with having a public API. With regression tests, you'll know that the features you deploy to production are, in theory, free from bugs that can potentially introduce outages and decrease customer satisfaction. Finally, regression tests also improve the usability of your API. Because your API has a low number of issues and behaves according to specifications, the experience of developers and other users interacting with your API will be the best possible.

API monitoring

So far, you've been learning about different ways of testing your API either manually or automatically as a part of a build pipeline. Running those tests periodically is also an important part of the QA process. With API monitoring, you'll be able to confirm that the quality of the API is the best possible at all times. Or, you'll know whenever something is not working as it should right away.

The first and most critical type of monitoring verifies something that has a deep impact on quality: the availability of your API. To put it simply, uptime monitors repeatedly check if an API is running and alert you whenever it isn't. Additionally, an uptime monitor can verify how long your API takes to respond to particular requests. For instance, you can set up an uptime monitor to make several requests and fail if the response times are above a certain limit. This possibility is handy if you have features that should respond within a specified time. Another interesting characteristic of uptime monitors is the ability to test the connectivity to your API from different geographies. By trying to access your API from disparate locations, you can assess how consumers from those places experience the API. Overall, an uptime monitor is the first line of defense you have against the misbehaviors of your API. This type of monitor can be set up using tools such as Uptime Robot or Postman.

The next type of monitor does what uptime monitors do for a more comprehensive analysis. While an uptime monitor checks the availability and response time of your API features, with a performance monitor, you can check how certain features behave with different types of request loads. A performance monitor can use the performance tests that you have previously configured and execute them periodically. It's important to understand that some performance tests can affect your API, so pay special attention if you're running the monitor against your production environment.

Another type of monitor that you should be careful with when running against a production environment is the security monitor. Since this monitor periodically executes security tests, running it on your live API can be dangerous if those tests are destructive. If, for example, some security tests purposely attempt to remove or manipulate information on the system, running them in a production environment can lead to disastrous consequences. I recommend removing those destructive tests from the monitor and running them against the production environment. The benefit of a security monitor is that you become aware that there's a security risk the moment it's detected. You can then act immediately by applying a security fix and preventing the problem from spreading.

Another type of monitor that is useful is what's called a usage monitor. By analyzing how consumers use your API, you can identify patterns and understand what areas of the API are used more often. Having a good knowledge of usage trends and patterns is critical to prioritizing new features and improving existing ones. You can analyze it passively by looking at usage logs and analytics. Or, you can be proactive about it and set alerts that are triggered whenever certain patterns appear. Tools such as New Relic let you set up this kind of alert that reacts to specific usage patterns.

In reality, any type of test can be transformed into a monitor. The main characteristic of a monitor is that it runs the test periodically and triggers an alert whenever the test fails. Depending on your business objectives, you will want to set up monitors to run different types of tests to maintain the quality of your API.

Summary

At this point, you know what QA is and, particularly, how it applies to API products. You also know that planning your tests is crucial to obtaining the best results from a QA process. You now know that behavioral and regression testing can be done by combining several test types identified earlier in this book. Finally, you know the importance of API monitors and how the information you obtain from them can help you maintain a high-quality API product.

You started this chapter by learning the definition of QA. Then, you learned how to plan the tests you run on a QA process to achieve your business objectives. You learned about different test environments and their purposes when related to several types of tests. Then, you went on to learn about behavioral testing, a way to combine several tests to learn how well the API is meeting the expectations of consumers. After, you learned about regression testing and its ability to detect if previously fixed defects are introduced by changes to your API. Finally, you learned that monitors are a way to periodically run several tests and alert you whenever they fail so that you can act quickly to achieve maximum quality.

Here are some of the aspects you discovered throughout this chapter:

- The notion of quality is hard to define

- A high-quality product has little or no friction

- QA measures how well your API product can meet the expectations of consumers

- The goal of a QA process is to confirm that the API aligns with business requirements, user stories, and needs

- Fixing defects before users notice them increases the quality of your API product

- Identifying the scope of your QA process is the first step to increasing the quality of your API

- There's a relationship between QA test scopes and environments

- Behavioral testing is a way to combine several tests to understand how the API behaves in different scenarios

- Boundary tests are useful for verifying how the API behaves in fringe scenarios
- Understanding different API states is crucial to knowing if things are working as expected
- Regression testing helps you identify if bugs that were fixed come back after you introduce changes
- Business stakeholders see regression testing as a way to minimize risks
- With API monitoring, you can automate existing tests
- Monitoring triggers an alert whenever a test fails
- By acting to mitigate problems when they happen, you increase the quality of your API

At this point, you know how API QA is related to tests and how important it is to guarantee that you have an API that meets consumers' needs. Monitoring is especially important because it automates your QA process and alerts you whenever something goes wrong. QA is mandatory if you're serious about the quality of your API product releases. In the next few chapters, you'll learn everything about the API release process. Keep reading to start learning about API deployment.

Part 4:
Releasing an API Product

This part of the book provides a comprehensive understanding of the latter stages of the API life cycle, beginning with an overview of the API deployment process encompassing continuous integration, agility, automated testing, deployment, and considerations regarding API gateways. It then explores API usage analytics, APM, and user feedback analysis for identifying and measuring essential metrics, usage patterns, and behavior. The part concludes with a thorough examination of API distribution strategies, including pricing, API portals, marketplace listing, and documentation options, all geared toward optimizing user activation.

In this part, you'll find the following chapters:

- *Chapter 13, Deploying the API*
- *Chapter 14, Observing API Behavior*
- *Chapter 15, Distribution Channels*

13

Deploying the API

Previously, you learned how you can use testing to increase the quality of your API product. Testing helps you find a better alignment with the needs of your users. You also learned how to set up a system that can run your tests automatically on every deployment you do to production. And that's how we're starting this chapter. One aspect of deploying an API is being able to automate tests, but there's more to it. Keep reading to learn all about it.

You'll start this chapter by learning what **continuous integration** (**CI**) is and what its role is in the task of deploying an API to production. You'll learn that there are automated processes that you can use to help you make your API available to consumers. You'll then get to know the difference between continuous delivery and continuous deployment. While they sound similar, they're different in the nature of what you can automate. You'll then learn the role of a **version control system** (**VCS**) in the process of automating the API deployment. After that, you'll see how crucial it is to have an automated build system that you can use to generate an SDK and manage configuration and dependencies, among other things. Right after, you'll see how API versioning plays a significant role in the deployment process. You'll learn how API versions can be communicated between consumers and your backend. Then, you'll learn about incremental, semantical, and calendar-based versioning. Near the end of the chapter, you'll learn what an API gateway is and what role it plays in the way consumers interact with your API. You'll also learn what are the most important factors to consider when choosing which API gateway to use. Finally, you'll see the difference between choosing a fully managed cloud API gateway and an open source one.

By the end of the chapter, you'll be able to understand what it takes to make your API available in a production environment. More than that, you'll understand what are the methods and tools that help you automate the deployment process. You'll know what it takes to have a CI process running to automate tasks behind making your API available to consumers. You'll also know how API versioning is a crucial part of deploying an API. Finally, you'll be able to choose an API gateway for your product.

In this chapter, you'll learn about these topics:

- CI

- Choosing an API gateway

Continuous integration

Continuous integration, or CI, is a methodology that involves the integration of multiple product changes into a common repository. Those changes are often produced by diverse people and teams. The goal of CI is to regularly identify any changes to the product and attempt to generate a new version of the product that incorporates all the latest improvements. CI uses automated processes heavily to reduce the friction of the process. In particular, automated testing and building the software behind the API server are two key components of CI.

Even though the origins of CI are not fully established, some people believe that it has evolved due to the popularity of the Agile software development movement. In particular, one of the creators of **Extreme Programming** (**XP**), Kent Beck, is known to be the one who coined the term "continuous integration." His goal with CI was to foster collaboration, adaptability, and quick feedback in software development. CI depends on common activities that are present in the software development life cycle. From continuous delivery and deployment to monitoring and alerting, CI touches all aspects of modern software development.

Let's start by exploring how continuous delivery and deployment work. To begin with, there's a big difference between continuous delivery and continuous deployment. Even though both activities imply a software deployment to a production environment, they're not the same. While the former offers a way to manually deploy software to production, the latter automates the activity. With continuous delivery, you get to manually review what you're about to make available in a production environment before actually committing to make the software available. On the other hand, continuous deployment takes the same process and automates it. The big difference here is that the automated process needs to offer a higher degree of trust because there's no human review. Usually, you add automated testing to a continuous deployment process. As you've seen in the previous chapter, if an automated testing process fails, the whole deployment to production doesn't happen.

One thing that is crucial to any CI system is integration with a VCS. A VCS, such as Git, can track any changes that you make to your code base and compare the code between several deployments to production. You can easily obtain just a list of changes that appeared since the last time you did a deployment to production. Furthermore, you can also track your machine-readable API definition document on a VCS. So, now you not only have your code versioned, but you're also doing it for your API definition. There's a big deal about this because you can use your machine-readable API definition to generate an SDK that clients can use to consume your API.

And, you can even automate the process and add it to the deployment to production. There are several tools that can help you generate an SDK easily. One example is the open source `openapi-generator` tool (go to `https://openapi-generator.tech` for more information). If you don't like installing software, you can easily use something such as Speakeasy to automate the task of generating an SDK. Both solutions take a machine-readable definition document and convert it into a ready-to-use SDK that clients can use to consume your API. Generating SDKs is just one of the things you can do during a build. You can also add automated documentation, or the ability to generate documentation for your API every time there are changes.

Actually, build automation is the most impactful CI activity. It's during an automated build that most things that are crucial to a successful deployment happen. Here are some of those things:

- **Compilation**: The act of converting code into files that you can run directly on the machine. In the case of API definitions, compilation can turn a machine-readable document into server code, an SDK, and documentation, among other things.

- **Dependency management**: Checking all the libraries and other artifacts that your project depends on. The list of dependencies can grow as a software project increases its complexity. Dependency management involves installing, updating, and tracking information about all the dependencies of the project.

- **Configuration management**: Setting up different environments and the configuration of how the API will run on each one of them. By using environment variables, infrastructure configuration scripts, and similar tools, you can effectively prepare your API to run in varied environments.

- **Quality assurance (QA)**: All the automated testing, validations, and reporting happen at this stage. If the automated QA process fails, the deployment process doesn't happen.

- **Build records**: Creating a receipt of the build and the deployment, including its result and information about any assets built during the process. Having build records is fundamental to enable the possibility of reverting the deployment to a previous version if needed.

Having a comprehensive automated build and deployment system accelerates the management of your API and reduces the workload of engineering teams. Another area that helps engineering teams and also increases visibility to your consumers is API versioning. Let's see in detail how it works and what you can do with it.

API versioning

The primary goal of versioning your API is signaling to its consumers that you introduced changes. From a consumer perspective, knowing what version an API is at is important. Because different versions of an API can have different features, consumers need to know which version they're interacting with. Sharing information about changes between versions gives API consumers the ability to decide what they need to update on their code base to use your newest version. Additionally, versioning an API helps you flag security issues and introduce fixes in a fluid fashion.

Ultimately, consumers should be able to choose which version of API to use in a simple way. Let's see what strategies you can follow to bake the versioning information into the way consumers interact with your API. All versioning strategies involve letting consumers send information about the version they want on every interaction with the API. There are different options that consumers can follow to identify the version of the API they want to interact with. The following list shows a summary of the most popular options and the API architectural types that consumers can use:

- **URL path**: Consumers indicate the version of the API they want to use by including it on the URL path; for example, `/api/v1`. You can use this option with REST and even with GraphQL, although it's not a conventional approach.

- **HTTP query parameter**: In situations where the URL path can't be directly manipulated, consumers can use a specific query parameter to indicate which version they want to use; for example, `/payments/version=1`. This option is often used with REST APIs.

- **HTTP version header**: API users add a specific header that holds the version information; for example, `X-API-version: 1.2.3`. You can implement this approach using REST and gRPC.

- **HTTP Accept header**: You normally use the HTTP `Accept` header to tell the API server what content type you want to consume. You can also use it to indicate which version of the API you're interested in. An example is `Accept: application/vnd.example.api.v1+json`. The content type is still JSON. However, there's extra information before the plus sign, including the version number. This approach is common in a REST environment.

- **Message topic or channel**: In an asynchronous environment, you can add the version you're interested in to the name of the topic; for, `vendor.payments.2023-01-10.event.payment.accepted`. This approach works well with **Advanced Message Queuing Protocol (AMQP)** or **Message Queuing Telemetry Transport (MQTT)** APIs.

These are some of the options you have for signaling the version of the API that you're interested in. Now, let's look at how you can define the version of the API. If you've paid attention, you'll note that in the previous examples, there were various ways of defining versions. The three most popular approaches are incremental versioning, semantic versioning, and calendar-based versioning. Let's see in detail what they look like.

Incremental API versioning

With the incremental versioning approach, you use an integer value to identify your API version. Whenever you introduce changes, you increment the version number. It's a simple versioning approach that's straightforward to understand and easy to use. Because of its simplicity, the version number can be added by consumers using any of the approaches you've seen before. As an example, consider a REST API with the `/payments` endpoint that exposes a resource collection. A consumer can, for instance, identify which version of the API to use by adding a version parameter. In this case, the consumer would make a request to `/payments?version=1`.

Because it's so simple, you can use the incremental API versioning approach using any of the options mentioned before. You can add it to the URL path or pass it as a part of an AMQP message title. However, the simplicity of the incremental version approach has the drawback of not providing much information to consumers. Other than the version number, you don't know why the API has a new version. That's where the semantic API versioning approach can help.

Semantic API versioning

Semantic versioning, or in short, SemVer, is an approach that lets you identify three types of changes on each version. This versioning approach was created by Tom Preston-Werner, one of the co-founders of GitHub, in 2010. Tom created SemVer to offer a standardized versioning scheme that conveys meaning about the nature of the change. Both humans and software can easily understand SemVer, making it ideal to use with API versions. More information about SemVer is available at `semver.org`.

SemVer can convey meaning to changes by breaking the version identifier into three components. The components are separated by a dot, and each one has its own meaning. Going from left to right, the first component is called `MAJOR`, and it holds the major version number of the API. You increment a major version whenever you introduce a breaking change; that is, a change that isn't backward compatible. Next, you have the `MINOR` component, which you increment whenever you introduce a non-breaking change. Non-breaking changes are those that are incremental or introduce something totally new. Finally, you have the `PATCH` component, which you increment whenever you introduce a bug fix that doesn't create a breaking change. Let's look at a few examples:

- **Moving from 1.2.0 to 1.2.1**: Because there was an increment in the `PATCH` component, it means that there was a bug fix

- **Moving from 1.2.1 to 1.3.0**: There was an increment in the `MINOR` component, which means that new functionality was added in a way that doesn't introduce a breaking change

- **Moving from 1.3.0 to 2.0.0**: This time, the `MAJOR` component was incremented, meaning that there was a breaking change

With SemVer, you can even convey information about prereleases and build information. This is an optional feature that you can use to convey more meaning. To inform consumers about prereleases, you add it after the `PATCH` component. The name of the prerelease can be anything you wish, as long as it's self-explanatory. As an example, you'd use `1.0.0.beta` to indicate that you're on the "beta" prerelease of version 1.0.0. To add build information, such as the date of the last build, you add a plus sign to the end of the version identifier. Examples include `1.0.0.beta+20230110` to indicate that the last build of the "beta" prerelease was on October 1, 2023, and `1.3.0+20230110` to indicate the same date for the last build of version 1.3.0.

Even though you can convey a great deal of meaning by using SemVer, sometimes all you really want is to let consumers know when you last updated your API. In those situations, what's important is to use the date of the last build as your version identifier. Keep reading to see how that works.

Calendar-based API versioning

With calendar-based versioning, you're directly sharing with consumers the date when you introduced a change to your API. This type of versioning works by using a date as the version identifier and following whatever approach you prefer to let consumers indicate which version they want to interact with. Using calendar-based versioning has the advantages of being simple to use and understand and making it easy for consumers to know how old the API is.

To use it, you pick the date when you introduced changes, and that becomes the version identifier. To follow a known format, write the dates using the *ISO 8601* standard. This international standard specifies a date format of "YYYY-MM-DD". As an example, you'd write the date of October 1, 2023, as "2023-10-01". Then, you let API consumers specify which API version they want by sharing the date on each request. The preferred method for sharing the date is using a dedicated HTTP header; for example, "X-API-version". If consumers don't use that header, it means they want to interact with the latest version of the API. In other words, they don't care which version they're consuming.

One thing that's between your API and its consumers is a gateway. While you can build an API without having a gateway, using one will certainly make things easier for you in the long run. Continue reading to learn how an API gateway can help you run your API.

Choosing an API gateway

API gateways are tools that have the primary responsibility of receiving requests from consumers and responding in a way that's aligned with your business objectives. From a consumer perspective, an API gateway is indistinguishable from the API itself. The API gateway receives requests from consumers and directs those requests to the real API. The gateway sits between consumers and the API backend that holds the code that powers the features:

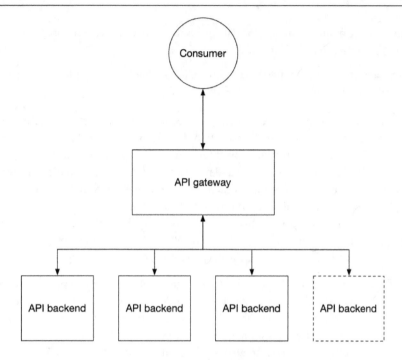

Figure 13.1 – Simplified API gateway architecture

In addition to being the bridge between the API code and its consumers, gateways can do a lot more. Here are the most important elements to consider when choosing an API gateway:

- **Architectural-type support**: Providing gateway capabilities to different architectural types, such as REST, GraphQL, gRPC, and asynchronous APIs

- **Request routing**: Determining which backend service should handle a particular API request based on the request's endpoint, HTTP method, or other criteria

- **Load balancing**: Distributing incoming requests across multiple instances or nodes of a service to ensure **high availability (HA)** and optimal performance

- **Authentication and authorization**: Enforcing security measures by verifying the identity of clients, ensuring they have the necessary permissions to access specific resources, and handling user authentication

- **Rate limiting and throttling**: Controlling the rate at which clients can make requests to prevent abuse or overuse of resources

- **Request and response transformation**: Modifying the structure or content of requests and responses to match the expectations of clients or backend services

- **Caching**: Storing frequently requested data or responses to improve response times and reduce load on backend services

- **Logging**: Capturing and analyzing API traffic data for troubleshooting, performance optimization, and security monitoring

So, depending on your business objectives, you might need to have one or more of these API gateway features available. Fortunately, the market offers solutions for all types of needs. You can pick a minimum API gateway or go with a fully featured one. Kong, for example, is a popular open source gateway that lets you start with a minimal configuration and add more features over time. If your company is already using a cloud provider, you can pick solutions such as **Amazon Web Services' (AWS')** Amazon API Gateway, or Apigee. Both gateways offer rate limiting, authentication, and security-related features out of the box. The big difference between solutions such as Kong and the ones offered by cloud providers is in the amount of work you'll have to spend time on. While cloud providers offer a fully managed solution, with open source solutions, you're the one configuring and managing the gateway.

An additional factor to consider is the cost of the API gateway solution you pick. Let's go back to *Chapter 3*, where you learned about API monetization models. Your goal is to align your choice of gateway with the way you're monetizing your API. So, if you're pricing your API using a tiered model, you want your gateway price to follow a similar, subscription-based model. On the other hand, if you have a pay-as-you-go model, you'd prefer to also pay for your gateway based on how much you use it.

Summary

By now, you have a good understanding of what it takes to deploy your API to a production environment. You can understand how a CI process can make your deployment fully automated. You also know the different types of API versioning and how they can convey different information to consumers. Finally, you know how various types of API gateways line up and what factors to take into account when deciding which one to use.

You began by exploring the definition of CI and how it became a crucial element of any API deployment operation. You learned about the different components of a CI process, such as compilation, dependency and configuration management, QA, and logging. You then got to know how API versioning is the central pillar of CI. You learned about incremental, semantic, and calendar-based versioning. You also learned what are the ways consumers can follow to request a specific version of your API. Then, you learned what API gateways are and how they can help you control the way consumers interact with your API. Finally, you got to know the factors that help you choose which type of gateway is better for your business.

These are some of the things that you learned during this chapter:

- CI integrates product changes into a common repository

- CI became popular with the growth of the Agile software development movement

- Continuous delivery involves manually deploying your API to production

- Continuous deployment automates the deployment of your API

- A VCS is crucial to the success of a CI process

- You can use your machine-readable API definition to generate an SDK that consumers use to interact with your API

- Generating an SDK is one of the things that you can do with an automated build system

- Automated build systems can also help you manage your API's dependencies and their configuration on multiple environments

- API versioning is a way to signal to consumers that you introduced changes to your API

- Consumers can request a specific API version by embedding it on a URL path, an HTTP query parameter, a header, or an asynchronous message topic or channel

- Incrementing the version number of your API is the simplest approach to communicating changes

- With semantic API versioning, you can communicate breaking and non-breaking changes and bug fixes

- Calendar-based versioning gives you the power to convey the date when you introduced changes to your API

- An API gateway is a piece of software and infrastructure that mediates communication between consumers and your API backend

- There are API gateways for different architectural types such as REST, GraphQL, gRPC, and asynchronous APIs

- API gateways can provide features such as request routing, load balancing, authentication and authorization, rate limiting, and caching

- There are open source API gateways and also fully managed cloud-based solutions

At this point, you know what it takes to deploy your API to production. You understand the importance of working with a VCS such as Git. Versioning an API is crucial to the management of changes to enable a continuous deployment process. You also know that gateways are a fundamental piece of an API to reduce the workload of your team. Without a deployment to production, consumers won't be able to interact with your API. While getting consumers is important, it's even more important to understand how they're using your API. And that's exactly what you'll learn in the next chapter. Continue reading to get to know how to observe API behavior.

14

Observing API Behavior

Putting your API product in front of potential users is crucial to finding success. What's even more important is understanding how consumers use your API, their concerns, and how their usage affects your infrastructure. Continue reading to dig deeper into each one of these topics.

This chapter starts by introducing you to the world of API usage analytics. You'll learn that with analytics, you can obtain insights into how users are consuming your API. You'll get to know important metrics such as time to first request, usage volume, and traffic patterns. You'll then learn how you can break down metrics to obtain valuable business information, such as the number of monthly active users and information related to infrastructure costs. After that, you'll understand what application performance monitoring is and how you can use it to increase the quality of your API. You'll then learn how to monitor and categorize error responses to better understand what areas of your API need to be improved. After that, you'll see how you can maintain the reliability of your systems by using a dependency map that shows what third-party APIs you're consuming. Finally, you'll see how user feedback is a component that, together with analytics, can help you reduce the time required to mitigate issues. You'll learn how to build your API in a way that encourages users to provide feedback. You'll then see how you can aggregate and categorize the feedback you receive. By the end of the chapter, you'll learn how to scale a user feedback system, ensuring that engagement isn't lost in the process.

After you read this chapter, you'll understand how API usage analytics can be an ally to your efforts to maintain high quality. You'll also be able to use analytics together with user feedback to gain knowledge about your API from a consumer perspective. In the end, you'll understand which tools to use to capture usage analytics, and how to monitor performance, and how to provide the best possible user feedback system.

The following are the topics that you'll learn about during this chapter:

- API usage analytics
- Application performance monitoring
- User feedback

API usage analytics

API usage analytics provide valuable insights into how your API is being used, who is using it, and how it's performing. The most important metric to consider when studying the behavior of your API users is the time to first request. This metric measures the time a user takes from signing up to making their first request to your API. It's an important metric because it measures the friction that consumers have when they first interact with your API. Having a high time to first request metric means that consumers are finding it difficult to understanding how to use your API. So, your goal is to lower this metric as much as possible. You can do that by implementing the learnings from *Chapters 2, 5*, and *12*.

After consumers make their first request, there are other metrics to consider that will provide different insights. You can understand how consumers interact with your API by analyzing specific metrics such as usage volume, request rate, and traffic patterns. Let's look at some of the most interesting types of metrics you can use to learn how users consume your API.

Usage volume is a metric that tracks the total number of API requests over a specific period. It helps you understand the overall load on your API and identify trends in usage. You can, for instance, understand if your API is used more on weekdays or weekends. Another (often associated) metric is the request rate. This metric breaks down the usage volume into requests per second or minute. With this information, you can monitor traffic spikes and plan for scalability. Following this type of analysis, you can also investigate when the API usage is highest or lowest during the day, week, or month by analyzing traffic patterns. With this information, you can better allocate resources and decide when to schedule maintenance activities.

Let's now move to metrics that help you identify API consumers. The first metric to analyze is user identification. By identifying the unique users that interact with your API, you can understand adoption, engagement, and even potential misuse. You can also use this metric to calculate the number of active users per period. This allows you to obtain important product-related metrics such as **monthly active users (MAU)**. Knowing your API's number of active users gives you information about its retention, and knowing your API retention is important if you're following one of the monetization strategies from *Chapter 3*.

Another metric that's important if you're monetizing your API is rate limiting. By monitoring rate limiting and throttling information, you can understand how many consumers are making requests with a frequency higher than they're allowed. This can open monetization opportunities. Additionally, tracking cost-related metrics, including data transfer and infrastructure expenses, is essential for financial planning and aligning your monetization strategy.

While these are the metrics I consider the most important, there are others that you should keep an eye on. Here are some less common API metrics to explore:

- **Geographic distribution**: Understanding where your API traffic originates can help you optimize the delivery of your services. It may also be relevant for compliance with regional data regulations.

- **Endpoint usage**: Tracking which API endpoints are most frequently accessed helps you prioritize optimization efforts and identify which features are most popular.

- **Response time**: Monitoring response times for API requests helps ensure optimal performance. You can identify slow endpoints and potential bottlenecks.

- **Error rates**: Analyzing error rates and types of errors helps pinpoint issues in your API and allows for proactive problem-solving. It's crucial for maintaining a high-quality API.

- **API version usage**: Tracking which versions of your API are in use can help manage version transitions, identify deprecated versions that need to be retired, and ensure backward compatibility.

- **API performance metrics**: Metrics related to CPU usage, memory consumption, and other system-level performance indicators can help ensure that your API infrastructure is healthy.

These API usage analytics provide a comprehensive view of how your API is performing and how it's being used by consumers. By regularly monitoring these metrics, you can make informed decisions about API design improvements, security, optimization, and scaling. You can learn even more about optimization and scaling with another type of analytic that tracks how well your API is behaving. Keep reading to learn more about **application performance monitoring** (APM).

Application performance monitoring

API APM is a set of practices and tools that focus on tracking, analyzing, and optimizing the performance of APIs. Using API APM helps you ensure that your API delivers a smooth, efficient, and reliable experience for consumers. Read on to learn about the critical topics of API APM, including API response time, error monitoring, end-to-end transaction tracing, and dependency tracking.

The response time of an API is a fundamental metric in API APM. It measures the time it takes for an API to respond to a request. Consumers have little patience for slow or unresponsive APIs. Therefore, monitoring and optimizing response time is crucial to ensuring that your API meets user expectations. Most APM tools, such as New Relic or Dynatrace, let you obtain a report of the response times of all the operations your API provides.

One of the key aspects of response time monitoring is setting response time thresholds. By defining acceptable response time limits for various API endpoints, you can create alerting mechanisms that trigger notifications when response times exceed those thresholds. These alerts enable rapid response to performance issues, reducing downtime and preventing user dissatisfaction. This approach is similar to what you've seen in *Chapter 12* when reading about API monitoring. However, instead of periodically testing the API and measuring its response time, API APM is constantly calculating response times based on the requests that happen in real time.

An in-depth analysis of API response time data often reveals performance bottlenecks that need attention. These bottlenecks may be due to factors such as inefficient database queries, excessive server processing time, or network latency. By identifying and addressing these issues, you can optimize your API's speed and efficiency, ensuring that your API has the best possible quality.

Another area that API APM is helpful in is error monitoring. API APM tools can identify whenever a request generates a known error and report that information in different ways:

- **Error rate**: This metric measures the frequency of errors occurring over a specific time frame. A sudden increase in the error rate can serve as an early warning signal of potential issues that require immediate investigation.

- **Error types**: Errors are categorized based on their nature, such as authentication errors, data validation errors, and server errors. Categorizing errors allows you to identify common patterns and issues.

- **Error patterns**: Monitoring error patterns helps you detect recurring issues, such as specific endpoints consistently producing errors or errors associated with a particular user group. Identifying these patterns facilitates targeted problem-solving.

By consistently reviewing error data and categorizing errors, you can proactively address error-prone areas of your API. Similarly to response time monitoring, setting up alerts to trigger whenever the error rate exceeds a specified threshold increases the quality of your API.

Knowing about response times and error rates is crucial to offering an API that meets the expectations of consumers. Understanding what it takes to deliver a response is fundamental to improving the efficiency of your API backend. With end-to-end transaction tracing, you can learn how an API request initiated by a consumer is accepted by your servers and how it's processed internally until a response is produced and sent back to the consumer. Most modern API APM tools offer end-to-end transaction analysis that includes elements such as transaction visualization, response time breakdown, and information about external dependencies.

End-to-end transaction visualization gives you access to a visual representation of the API transaction flow, showcasing consumer interactions, API endpoints, backend services, and external dependencies. This visual map aids in understanding the holistic transaction flow, making it easier to identify bottlenecks and inefficiencies.

Response time breakdown extends the information on the time it takes to respond to requests by providing information on the time spent in each stage of the transaction. With this information, you can identify which phases of the request contribute most significantly to response time.

The final area, dependency tracking, is probably the most interesting to explore. Being able to identify – and monitor – external dependencies means that you can fully understand how your backend system consumes third-party APIs. A good example is when your API uses a payment service provided by a third party. Whenever your API processes a payment, it needs to interact with the external payment API, which becomes a dependency. While leaning on dependencies such as the one in this example helps you accelerate the development of your API product, you'll also become more exposed to points of failure that are out of your control.

Most API APM products can track, analyze, and map the web of external dependencies that your API uses. The first component is a dependency map that lets you visualize how your API connects and consumes third-party APIs. This kind of map is useful as a reporting tool, but it can also aid in debugging activities. With a dependency map, you can quickly understand the relationship between your system and external dependencies.

Additionally, to understand what connections your API makes to third parties, it's crucial to know if any external services aren't operating normally. This type of analysis is called dependency health and performance monitoring. Essentially, it measures the health and performance of any third-party dependencies and exposes that information to you. It works by monitoring all external dependencies and measuring their response times, error rates, and uptime. In a way similar to what is possible with other monitoring tools, you can set up alerts that trigger whenever any measurements fall outside of your defined thresholds.

In addition to mapping and understanding the health of external systems, API APM tools can also calculate the impact generated by dependencies. While some third-party dependencies are critical to your API, others have a lower impact. Understanding the overall impact of dependencies helps you assess risks and proactively apply mitigation strategies, such as switching providers, if needed.

As you can see, effective API APM not only ensures the reliability and scalability of your API but also contributes to higher quality, leading to a higher chance of business success. While API APM is something you can set up and run internally, there's nothing like understanding what your consumers are experiencing. Keep reading to learn how you can obtain and measure user feedback to enhance the quality of your API.

User feedback

API consumers can provide something that you can't get from anywhere else: their unbiased opinions about your API product. Whenever consumers find performance issues, errors, or inefficiencies, they are often the first to notice, making their feedback a valuable source of data for observability. Additionally, they can report issues that might not be immediately evident through any of the quantitative analyses you saw earlier in the chapter. Consumers, for example, are able to share problems that are related to usability and functional challenges.

How you gather feedback from users is a critical aspect of analyzing what they have to share. The mechanisms you use to let consumers share their feedback should be seamlessly integrated into your product or, even better, into consumers' workflows. The first feedback channel you can think of is to open a direct way for reporting errors. To integrate that option into your API product, you can add information to error responses to show consumers how they can report issues. One way to do that is to follow the *Problem Details for HTTP APIs* proposed standard described in RFC 9457 (see `https://datatracker.ietf.org/doc/html/rfc9457`).

Here's an example of what an error response following this approach looks like:

```
HTTP/1.1 502 Bad Gateway
Content-Type: application/problem+json
Content-Language: en


{
  "type": "https://example.com/errors/gateway-unreachable",
  "status": "502",
  "title": "Unreachable payment gateway.",
  "detail": "We couldn't process the payment because the gateway is
unreachable. To report this issue please go to https://example.com/
support",
  "instance": "/payments/50a9249c-eb80-41fd-aa96-aa9d88b43f0c"
}
```

In this example, the API server couldn't reach the payment gateway and couldn't process a payment. Notice how the error response mentions how a user can report the issue. The consumer has the opportunity to contact support to provide more information related to the context of the error.

Another way to obtain feedback from consumers is to establish various contact channels that they can use to access the support team. Remember that the goal is to use channels that consumers are familiar with. So, you could use a chat system such as Slack if you know that there's a large portion of users that are familiar with it. It's preferable to get as close to consumers as possible instead of making them come to you. Here are a few different options for support channels:

- Chat systems such as Slack, Discord, and Microsoft Teams
- Bug reporting systems such as GitHub and Jira
- Branded chat support systems such as Intercom
- Email
- Video or voice calls

Up to now, you've seen reactive feedback mechanisms. Consumers share their feedback whenever something doesn't work as they were expecting. However, there's another type of feedback that helps you proactively learn how consumers perceive your product. Conducting a user survey is one way to gather structured feedback on topics that you're interested in. You can structure the survey with the questions that you want answers to and learn from your users before the problems happen.

Aggregating and categorizing feedback

The diversity of user feedback demands a structured approach to categorization. It's crucial to organize feedback into meaningful categories that make it easy to identify trends and prioritize improvements effectively. Common categories for feedback include the following:

- **Performance issues**: This is feedback related to API response time, latency, or general sluggishness. Consumers might report instances of slow performance, which could be an early indication of underlying issues.

- **Functional problems**: These are reports of errors, broken features, or functionalities that do not work as expected. This category includes issues such as incorrect data, failed transactions, or unresponsive endpoints.

- **Feature requests**: These are user suggestions for new features or improvements to existing functionalities. These requests often provide valuable insights into consumers' needs and expectations.

- **General comments**: These are unsolicited comments, user testimonials, or expressions of satisfaction. While not problem-specific, these comments can offer a holistic view of user sentiment.

If categorizing feedback helps in creating a structured feedback repository, aggregating it offers insights into how consumers perceive your API product. To begin with, you'll be able to analyze the trends in the issues that consumers are reporting. You can then prioritize the issues and requests that make the most sense to you and your customers. Another advantage of having a repository of aggregated feedback is accessing historical context. Whenever users interact with you, you'll be able to understand their past interactions and how their requests were fulfilled. Additionally, if users know that you care about their past interactions, they're more likely to participate in the future and will have a more positive opinion about your API product.

All this sounds great, but unfortunately, it takes time to put together and requires lots of human labor. That is, unless you use the power of automation to analyze and aggregate all the feedback you receive. **Natural language processing (NLP)**, **artificial intelligence (AI)**, and sentiment analysis provide a solution. By using these technologies, you can automatically categorize and assess the sentiment of user feedback, making the feedback analysis process more efficient.

NLP and AI can categorize feedback by identifying keywords and phrases associated with specific issues. For instance, if a user mentions a slow response time in their feedback, the system can categorize the feedback as being related to performance. Sentiment analysis can determine whether user comments are positive, negative, or neutral, helping you identify areas that require immediate attention.

Another way of categorizing feedback is to use quantitative metrics. These are direct measurements of the feedback and the overall satisfaction of users. Here are the most important metrics that you should consider:

- **Number of feedback submissions**: You can track the volume of feedback submissions over time. A sudden increase in submissions can indicate a recent issue or concern.

- **Feedback response times**: Measure the time it takes to respond to user feedback. Quick response times demonstrate attentiveness and a commitment to resolving user concerns.

- **User satisfaction scores**: Implement satisfaction surveys that allow users to rate their experience with the API. These scores provide a direct measure of user sentiment. Popular scoring systems include the **net promoter score** (**NPS**), the **customer satisfaction score** (**CSAT**), and the **Customer Effort Score** (**CES**).

Acting on feedback

After you have a system in place to capture, aggregate, categorize, and measure the effectiveness of your feedback, it's time to make the information actionable. It's clear that there are types of feedback that are related to critical or urgent situations, and other types don't require immediate attention. A good approach to make feedback actionable is to implement alerts that trigger whenever the feedback you receive requires instant care. Alerts can be based on various criteria, such as the severity of the issue, the number of users affected, or the potential impact on the API's performance. It's up to you to understand and implement what makes the most sense in your particular situation.

For example, if a significant number of consumers report a specific issue, such as an inability to complete a payment, an alert can notify the development team. This enables rapid response to critical problems, helping to minimize disruption and maintain user satisfaction.

While user feedback is invaluable, it becomes even more powerful when combined with technical observability data. Integrating user feedback mechanisms with existing observability tools, such as APM platforms and logging systems, lets you correlate user-reported issues with technical data. In the previous example, if there's an integration with APM, your development team can understand the technical reasons why the payments are not working and correlate the information with the feedback from actual API consumers. This correlation facilitates more accurate issue diagnosis and resolution because it often offers clues that would otherwise take longer to uncover.

One particular area where user feedback is valuable is performance. How fast your API operates directly affects the user experience, as you've seen in *Chapter 2*. API observability in tandem with user feedback can provide the resources you need to improve performance where it most affects the quality of your API. Performance impact analysis involves the following areas:

- **Root cause analysis**: This involves identifying the technical root causes of issues reported by users. This may involve tracing API calls, examining database queries, or assessing network performance.

- **Performance bottleneck identification**: This involves determining which technical aspects are responsible for performance degradation and whether they align with the user's reported concerns.

- **Impact assessment**: This involves understanding the extent to which reported issues affect the overall performance and functionality of the API. This assessment guides the prioritization of improvements.

User feedback serves as a valuable starting point for performance impact analysis. When you can connect user-reported issues to technical insights, you can make more informed decisions about issue resolution and optimization.

Scaling user feedback

As your API product grows and gains more consumers, the volume of user feedback can increase significantly. It's essential to ensure that feedback mechanisms and systems are scalable to accommodate this growth. Scaling a user feedback system involves automating the process of collecting information, scaling the aggregation system to handle large volumes of data, and maintaining a positive experience that avoids friction during the feedback process.

You can enable the automation of collecting feedback from users by using tools that can work with the least amount of human intervention. Services such as Zendesk or Intercom offer a semi-automated system that can even resolve issues with no human intervention at all. Whenever required, those systems will offload the communication to a human, usually someone working on user support. It's important to understand that the more you aggregate and categorize the feedback, the better the automation systems can perform.

And that's the second area that you'll need to scale. While aggregation and categorization can work easily with a relatively low amount of data, you need to keep up with the increase in the volume of information. Here, I recommend using an existing tool that offers scalability out of the box. If, for example, you're using Zendesk or Intercom, some of the aggregation and categorization features are already available for you. Scaling, in this situation, means having to pay more to use those products. However, it removes the burden of having to scale your own internal aggregation systems.

The third elements that require scaling are the interaction points with users. Here, your goal is to make sure that you don't degrade the user experience of providing feedback. If, in the beginning, you were manually responding to all the support requests, you now won't be able to handle the increasing volume of inquiries. But you need to maintain the same feeling of engagement that you offered in the beginning. Otherwise, users will feel discouraged to provide feedback, and their perception of your API product and company will eventually become negative.

Engagement is, in fact, the most important part of a positive user feedback system. From small to large user bases, effective communication is essential to demonstrate that the feedback you receive is valued and acted upon. These are a few of the elements that help you keep your use base engaged and open to providing feedback:

- **Transparency**: This means being transparent about how you use feedback, what changes are being made as a result, and how feedback contributes to ongoing improvements

- **Issue resolution updates**: This means informing users about the resolution of the issues they report, ensuring that they know their concerns have been addressed

- **User community engagement**: This involves creating opportunities for users to engage in discussions, forums, or community boards where they can share feedback and ideas

- **User-focused roadmaps**: This refers to sharing product and feature roadmaps that reflect user feedback and priorities

Overall, feedback is a valuable source of information and a way to keep your users engaged. When your user base is happy with how you respond to their feedback, the perception they have of your API product is positive.

Summary

At this point, you know what it means to observe API behavior. You can identify the different metrics that are important to consider while doing API usage analysis. You also learned that while it's important to understand user behavior, knowing what happens on your backend system is fundamental. You know how to use APM as your ally to understand how to mitigate problems even before users report them. Speaking of users, you also now know the importance of user feedback and how it can help you maintain a high-quality API product.

You started by learning all about API usage analytics, including the most important metrics to explore. You learned what time to first request is and how it reflects the user experience. You also learned about other metrics such as usage volume and patterns, number of errors, rate-limiting violations, and performance. You then got into the topic of application performance monitoring and learned how you can use it to capture metrics that are internal to your systems. You also learned how to set up alerts that are triggered whenever metrics reach certain thresholds. In the last section of the chapter, you learned about user feedback and how you can use it in conjunction with usage analytics and APM. You also learned how to aggregate and categorize feedback information to obtain meaningful insights. In the end, you got to know how to scale a feedback system to endure the growth of your API product without compromising user experience.

Here are a few of the topics you learned about in this chapter:

- The most important metric to consider is the time to first request
- Identifying unique API consumers lets you obtain product-related metrics such as the number of monthly active users
- Most APM tools give a report of the response times of all your API operations
- Setting thresholds on metrics lets you enable alerts
- End-to-end transaction analysis gives you information to identify bottlenecks
- Dependency tracking lets you learn how your API is using third-party services
- One of the ways to guide consumers to provide feedback is to follow the *Problem Details for HTTP APIs* proposed standard
- User feedback should happen in places that users are familiar with
- Aggregating and categorizing user feedback lets you understand trends.
- You can use NLP and AI to automatically categorize user feedback
- User satisfaction scores such as the NPS let you understand how happy users are with your feedback system
- You can scale user feedback by automating most of its parts

Right now, you have enough knowledge to understand what the different components of API behavior observability are. You understand the available tools and techniques that let you implement usage analytics, APM, and user feedback. Understanding how users interact with your API is crucial. Being able to put yourself in their shoes is even more important, and that can be done with the right combination of user feedback and analytics. However, having access to all these techniques without having users is meaningless. That's why you'll learn how to take advantage of distribution channels to establish and grow your user base in the next chapter. Keep reading to learn about the different options you have to get your API into the hands of consumers.

15

Distribution Channels

An API product doesn't make sense without consumers interacting with it regularly. Until now, you've been learning how to build your API in a way that's aligned with the expectations of consumers. Now, you'll see how you can make your API discoverable to the potential customers that you want to attract. Keep reading to learn how you can use distribution channels to deliver your API and grow your user base.

In this chapter, you'll begin by learning what exactly API distribution is and how it's related to the ability to generate revenue. You'll start by learning about the **go-to-market (GTM)** strategy, how it helps you identify the market where you want to operate, and how using customer personas helps you identify the positioning of your API product. Then, you'll learn about pricing strategies. You'll see that you can combine different monetization models in ways that make it easy for your audience to subscribe. You'll also learn that you need to have the right pricing in place to at least cover the costs of acquiring new customers. You'll then explore in detail different ways of making API distribution work. You'll learn the traits an API portal should have to become a viable distribution channel. After that, you'll learn how you can engage the community of consumers to help you effectively distribute your API. Then, you'll see what API marketplaces are and how you can use them to attract more customers. Finally, you'll see how documentation can be an interesting way to distribute your API product, by creating inbound channels.

By the end of the chapter, you'll be able to understand how a distribution strategy is crucial to attract customers to your API product. You'll know how to price your API so that your **customer acquisition cost (CAC)** is always lower than the **customer lifetime value (CLTV)**. Finally, you'll know how to leverage four distribution channels: API portal, community, marketplaces, and documentation.

This chapter covers the following topics:

- What is API distribution?
- Pricing strategy
- API portal
- Community

- Marketplaces
- Documentation

What is API distribution?

API product distribution is the process of making your API available to consumers in the easiest possible way for them. One key factor of API product distribution is understanding the tools consumers use and making your API available to them on those tools. Doing that guarantees that consumers won't have to change their existing workflows just to be able to use your API product. Successful companies understand that they need to make their products available where their customers are. A good example of a physical product that followed this approach is Coca-Cola. In its early days, the now-popular beverage was available in soda fountains at pharmacies. After all, in the beginning, they were selling a medicinal beverage. And where else should they be selling it other than pharmacies? Over time, Coca-Cola adapted its distribution model as the product itself evolved and became attractive to different types of customers.

As you've learned before, APIs are products. However, they're not physical, like a beverage is. But they also have a business value, a monetization strategy, and consumers. In *Chapter 2*, you learned that your API users can be spread across different industries and geographies. You also learned about the concept of API friction, and how to minimize it. Now, you'll explore ways to distribute your API to your target market so that they can start using it with the minimum amount of friction. In the end, your goal is to generate revenue from your API product.

One critical aspect of a distribution strategy is called GTM. A GTM strategy is a plan that guides you and your company into successfully launching your API product with the goal of generating revenue. You start by analyzing the market where you want to operate so that you can position your API in the best possible way. The goal of analyzing the market is to understand who your customers are so that you can make your API product attractive to them. It's during market analysis that you identify how big your target market is. As an example, for a payments API product, you can identify small and medium companies as your target market. Then, you can size the market by analyzing how many small and medium companies exist. You can further refine your market by segmenting it into chunks that share common elements. In the case of a payments API, you could segment your total market into countries, where each country has its own way of making and accepting payments.

After identifying your potential market, it's time to define your strategy for acquiring customers. Your first tool is the customer persona. With it, you can identify who your potential customers are and how you can provide something they find valuable. Then, you need to make the target market fully aware of your API product and how it can solve their problems. Your goal is to attract people that belong to your target market to your API product. The way to measure success is to count how many signups your API has over a period of time. Be aware that there's a cost associated with the acquisition of every customer. It's called the CAC. You should always aim at acquiring customers at a cost that is lower than the revenue you generate. Otherwise, you're acquiring customers at a loss.

For every new customer who signs up to use your API product, you should be able to generate revenue. You can do that by managing how consumers subscribe to your API following one of the monetization models introduced in *Chapter 3*. While the freemium and tiered models produce a predictable monthly revenue, the **pay-as-you-go** (**PAYG**) model offers a less consistent income, as it depends on how much consumers use your API. Let's now see in detail how you can use these and other pricing strategies in combination with how you distribute your API product.

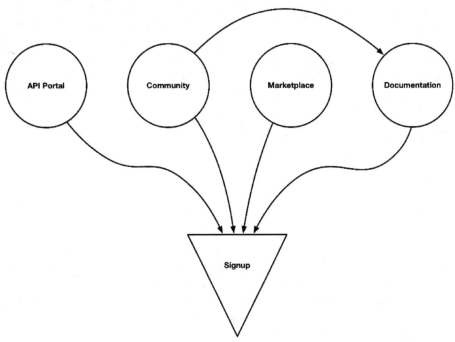

Figure 15.1 – A simplified view of several API distribution channels

Pricing strategy

Choosing the right pricing strategy for your API product is a big part of your road to success. Understanding your customers, the market where they're located, and their preferences in terms of buying is critical to defining the way you price your API. For example, small and medium-sized companies prefer an easy-to-pay subscription service, while larger-sized companies prefer to bind their cost to how much they consume.

In *Chapter 3*, you learned about the freemium, tiered, and PAYG monetization models. However, there are other models that you can use as an alternative or in combination with the three you learned about before. Let's start with the freemium model, which offers certain features for free while charging for others. The free features provide enough evidence of value for consumers to want to upgrade to the paid version. The freemium model can be used in combination with any other paid model. The Google Maps API, for example, follows a strategy that combines freemium with PAYG. In November 2023, Google offered 28,500 requests per month to its Maps API for free. It even went further and translated those free requests to the value of $200 to show potential users how much they could save every month. Any usage beyond the free limit was charged following the PAYG model. GitHub follows an alternative strategy where you only have access to certain features via the API while subscribing to the free tier, whereas consumers who pay can use more features. While the API itself can potentially access the whole range of available features, users can only access those features they pay for via the API.

Even the tiered model has many combinations that can be relevant to different types of customers. Most tiered models offer a monthly subscription. However, there are cases where offering a discount over a quarterly or annual upfront payment makes sense. By billing customers on an annual basis, you'll get their full payment immediately and guarantee that they'll stay with you for at least 1 year. This type of billing only makes sense in markets where your average customer has the necessary spending power. Otherwise, you'll end up in a situation where almost no customers will sign up for the annual subscription plan.

And even customers who can spend the full annual amount are sometimes reluctant to do so. Offering a discount is one way to convince customers to spend more upfront. Another way is to offer a trial proportional to the total subscribed period. So, in the case of an annual subscription, you offer a 30-day trial, and for a monthly subscription you offer a 7-day trial. The goal is to let customers try the product before they're billed. If, before the trial period ends, they change their mind, they can cancel the subscription.

There's an interesting alternative for situations where customers aren't willing to pay the subscription prices you're aiming at. It's a combination of the freemium model with the subscription of individual features. Essentially, you're letting customers build their own subscription plans by picking just the features they need. Each feature has its own subscription price and can be purchased individually or as a bundle of features. The freemium part of the model encourages customers to use the API product for free, even though it has visible limitations. Then, customers who want more features – or enhancements to the features they're using for free – can subscribe to them individually.

Something else to keep in mind regarding pricing is the customer's company size and the bureaucracy that might be involved in approving the payment of a subscription. With small and medium-sized companies, you can expect the approval of an expense to be relatively fast. You can also expect those customers to be able to pay a subscription using a credit card and to finish the signup and payment process on their own. On the other hand, larger companies tend to have heavy expense approval processes in place that not only make the first payment take long but also make it difficult to set up a recurring payment for a subscription. That is why many existing API products offer a much higher starting price for large companies. Usually, they refer to those customers as "enterprise" and the process involves at least one interaction with a sales team.

Your goal is to align your pricing strategy with your costs of running the API and acquiring customers. There's a useful tool called CLTV that you can use to understand how much value you can earn from each customer. Then, you can adjust your prices so that your costs are always below the CLTV. A simple way to calculate CLTV is to multiply the average revenue generated per retained customer by their estimated lifespan. So, as an example, for an annual subscription price of $100 and a retention rate of 75%, the average revenue per retained customer is $75. The estimated customer lifespan is calculated by inverting the churn and, in this example, is 4 years. The CLTV can then be calculated by multiplying $75 by 4 years, giving a total of $300. With this information, you know that your CAC has to be lower than $300. Otherwise, you'll have a loss because you're spending more per customer than what you're earning.

$$CLTV = \frac{ARPU \times r}{1 - r}$$

In the end, it all has to do with how customers perceive the value of your API product and how much they're willing to pay for a subscription. A big part of how perception works is by paying attention to the packaging and the onboarding of a product. If the customer experience is frictionless then the perceived value will be much higher than it would with an API product difficult to work with. One way to decrease that initial friction and make onboarding a pleasure is having a high-quality API portal. In *Chapter 3*, you learned the importance of a developer – or API – portal and how it's a part of your API support and documentation. Now, let's see how a portal can help you distribute your API product to your audience.

API portal

When you walk on a street filled with shops, your attention drifts to the storefronts that attract you the most. That's a natural reaction; as you're drawn to the things that you care about the most, you'll unconsciously look for them in the world around you. A similar thing happens in the digital world. Notice how you pay more attention to topics that you're interested in and, almost automatically, discard anything that isn't in your range of interests. This ability is called *selective attention*, and it's what lets you direct your mental resources toward particular stimuli while filtering out less important information. Top-down processing is one of the mechanisms of selective attention that uses your pre-existing knowledge and expectations to decide what to do with new information.

This mechanism is also responsible for immediately discarding anything that isn't aligned with your own belief system, including a storefront.

If you think of a portal as a storefront for your API product, its most important aspect is how it makes potential customers feel when they first see it. If those potential customers don't feel attracted to your API portal, they won't even explore what's behind it. So, your first task is aligning the look and feel of your API portal with the selective attention of your target audience. The branding and communication that you use on your API portal should then reflect how you want to appear to your potential customers. After all, you only have one chance of generating the right first impression.

Notice, for example, how the branding and messaging differ between popular API products such as Stripe, Twilio, and Algolia. Stripe shows you how it can become your payment processor – it even calls itself the "*Payments infrastructure for the internet*" – by presenting financial information such as sales charts, payment forms, and credit card brands. Twilio's focus is on presenting itself as an enterprise-grade communications platform by showing some of its features such as SMS fraud protection, and giving examples of large companies that are already customers. Finally, Algolia presents itself as a search platform that you can start using for free and easily integrate with your website. By adapting their branding to the style and expectations of their respective audiences, each of the previous companies was able to cement their position in each of their markets.

Since your goal is to put your API product in front of as many customers as you can, your focus should be to make it extremely simple for developers to build integrations with other products and solutions. So, instead of giving the highest importance to your API reference and to how a developer can access it, you should work on documenting use cases that solve the problems of potential customers. Write a tutorial on how to use your API for each of the identified use cases. Developers will then see value in those tutorials and will follow them to build the integrations they're looking for. Think about each use case not just for the problem it solves but also for how a developer would build the code to solve it. If the use case is related to a mobile interaction, offer ready-to-use code that developers can implement on their mobile apps. If the use case deals with a high volume of requests, offer a way to queue and monitor those requests. In summary, don't just document how a developer can use your API; show them how to implement the particular use case they're interested in.

Community

Being able to cover all the possible use cases that your audience is interested in is almost impossible. That's why you need to have a discovery mechanism in place and keep updating your API portal with any newly found use cases. One way to give developers the information they're looking for is by offering access to a community of like-minded people. You can, for example, invite developers to share their experience integrating with your API. That can materialize in the form of an article in your API portal and also as part of a forum where other developers can also participate. By following this approach, you'll be able to learn about use cases that you couldn't cover in the first place while giving developers a way to voice their needs to the community.

Amazon Web Services (AWS) is a good example of how this approach can be successful and you can see it as something to follow. There are different community forums where developers can participate, such as re:Post or the AWS Collective. With nearly 100,000 active developers and hundreds of thousands of topics in 2023, AWS has successfully built a community around its API products. Not only are developers able to share ideas and learn from peers, but AWS employees also gain access to a vast repository of community-generated knowledge. An AWS product manager can, for example, understand what popular questions of the week are and proactively engage with customers who might face similar challenges.

But why stop at offering an online place where your API users can congregate? You can also organize events, meetups, and hackathons and invite members of your developer community to meet each other and interact in real life. It's true that during the 2000-2022 pandemic, in-person events lost popularity. In those years, and especially in places where there was a mandatory lockdown, people opted for getting together online. However, in 2023, in-person events started to pick up again and have been gaining popularity ever since. Hackathons are one of the best types of events for putting new technology in the hands of developers. The term *hackathon* is a combination of the words *hack* and *marathon*. The goal of a hackathon is to work on a brand-new project or feature during a pre-defined period. The most popular format consists of a contest where several teams of developers – the hackers – compete for the best project. It's one of the best ways to promote a new API product, so you should use it whenever you can.

Another type of in-person event that's popular among API enthusiasts is a meetup. While it's an informal event that people join to meet like-minded peers, you can use it to your advantage. A meetup is the ideal scenario to promote a new API product or to encourage the discussion and presentation of ideas related to your API. It's also a way to give back to the community and associate your name with the topic of the meetup. A company that's been using meetups successfully is MuleSoft. With more than 100 meetup groups scattered across the world, in 2023, there were in-person events in more than 20 cities in countries such as Australia, India, Sweden, and the United States. The company encourages anyone to become a meetup leader to start a community in their city and even welcomes other companies to sponsor local events. This is an example of how being open to the community lets you expand your brand and have people all over the world engage with your API product.

Having a community is a way to expand your audience and create more opportunities for people to start using your API product. However, there are drawbacks to consider when attempting to scale the acquisition that a community can generate. First, you should think about the cost of acquiring each new customer. If the CAC gets too high, having a community stops being an interesting way of distributing your API product. On the other hand, if you manage to keep the CAC low enough, you might find yourself in a situation where you can't scale. You'll eventually reach a point where you'll reach a plateau in the number of new customers you can acquire through the community. This is when it's interesting to consider another option that you can add to your list of distribution channels: the API marketplace.

Marketplaces

A storefront is how you attract potential customers to your product. But to make people become real customers, you need to get enough of them to see your storefront and walk into your store. That is exactly what marketplaces do. In the physical world, a good example of a marketplace is a supermarket. The promise that a supermarket makes is that buyers can get everything they need in one single place. Instead of having to wander between different stores, possibly located distantly from one another, a buyer can simply walk into a supermarket. From a seller's perspective, supermarkets can draw more potential buyers than single stores would. That's exactly why marketplaces are an interesting distribution channel to explore.

In addition to promoting your API product by yourself, you can also make it available through API marketplaces to reach wider audiences. In 2023, the most popular API marketplaces included Rapid, Postman, and SwaggerHub. All these marketplaces have in common that they also let you design and manage your API, not just showcase it. And, in the case of Rapid, you also have monetization options available. Thinking exclusively about distribution, the differentiators you want to look at are related to the size of the audience available on each marketplace and how easy it is for you to reach it. Rapid, for instance, claims to have an audience of 4 million developers, while Postman shares that it has over 25 million users, and SwaggerHub users are just over 100,000. Measuring how easy – or hard – it is for you to reach your audience has to do with the total number of APIs available on each marketplace. Rapid and SwaggerHub are similar and claim to have approximately 40,000 APIs available to use. Postman uses workspaces, not APIs, in its listings, and claims to have over 200,000. In short, making yourself visible among thousands of other APIs can have its costs. But it also has many benefits. One of those benefits is related to discoverability.

Marketplaces attract API consumers for the convenience they offer during the process of discovering an API that solves their use case. Marketplaces work in a way similar to popular search engines. While they're not as sophisticated in terms of ranking as something like Google, they all offer search and discovery features. To be able to appear on search results, you need to understand how search and discovery work on the marketplace you choose to use. In essence, you need to know how to become the first result when users search for keywords that match what your API product offers. Make sure you use words in the description of your API that align with the use cases it solves and with how consumers will search. If the marketplace you decide to use accepts tags or categories, it's a good idea to use them. It helps users navigate and discover your API more easily than by just searching. Even with all these things in place, you might find it hard to make your API relevant enough so that users find it in a marketplace, especially if the marketplace is too crowded. One way of becoming more relevant is by becoming a featured API.

Most marketplaces offer featured listings, ads, or a combination of both. And you should take advantage of what is available even if, in some situations, you'll have to pay. Remember what you learned before about the cost of acquiring new customers. As long as the CAC is not higher than the CLTV, you can make use of paid tactics to increase the visibility of your API. The way to be featured in one of the marketplaces is to offer a high-quality API, one that the marketplace considers worthy of featuring.

Then, establish a close relationship with the company behind the marketplace and present your API and its advantages to consumers. This can take some time, depending on how crowded the marketplace is and how the company behind it handles the relationship with API owners. The result is that they can feature you and give a traffic boost to your API. You might argue that this tactic feels like gaming the system, and I wouldn't disagree with you. If you prefer to follow a more organic approach, there's another option that measures how popular an API is and bubbles it to the top of lists.

Encouraging users to interact with your API listing in the marketplace will increase the chances of it becoming popular. The difficulty of this approach is in being able to attract enough people to your marketplace listing. To do that, you can make use of your API portal and advertise your presence in the marketplace. You can also invite existing API consumers to the marketplace. Postman, for example, offers a button that you can easily add to any web page. When users click on the **Run in Postman** button, they're directed to your API entry on their marketplace. From there, they can directly interact with your API, and, by doing that, they're increasing its popularity. The more popular an API is, the more relevance it gets in search results, and, eventually, it will be featured in the marketplace.

Why would potential users want to use your API through a marketplace instead of consuming it directly from your API portal? That's a good question and one that deserves attention. One factor is related to convenience. If someone is already signed up on a marketplace and already uses one or more APIs, they'll prefer to consume a new API from the marketplace. This is because the marketplace offers a standardized approach to discover, consume, and debug APIs. Instead of having to learn everything from scratch, consumers can jump the initial hurdle of understanding how to use your API by following the standardized approach the marketplace offers. Another factor is related to the integration options the marketplace makes available.

Most API marketplaces offer a way to generate API client code – and even full SDKs – in several programming languages. Some of them let consumers work with APIs by using their favorite development tools. By being available in tools such as VS Code, marketplaces make it very easy for developers to interact with existing APIs. Potential API consumers could start exploring your API from within their code editor, import and generate client code, integrate the code into the app they're building, and finally deploy the integration. All from inside their favorite development environment. Altogether, the power of API marketplaces lies in their ability to offer a full end-to-end experience to consumers. This is something hard to replicate by yourself unless you already have a large audience and you can drive enough revenue to justify the high costs. By contrast, something else that you can and should do is build high-quality documentation that can be used on its own and also as a way to leverage your marketplace listings.

Documentation

Using documentation to attract new API consumers can be an effective distribution tactic if done correctly. One way to attract new users to your API portal is by making your documentation easy to discover. You can use what is called **search engine optimization**, or **SEO**, to make sure search engines process and deliver the content you write as search results. The way to do it is to think about your documentation from the perspective of potential users searching for a solution to their problems. The higher your content shows on a search results page, the more chances you have of someone opening your portal and, from there, signing up to use your API. So, as an example, if you're building a payments API, you'll want to focus on writing tutorials explaining how to implement a payment solution. Instead of focusing your documentation effort on explaining what your API does, you show users how they can implement a solution to their job to be done.

The content needs to be engaging enough so that users keep reading it and feel interested in trying the solution that you're presenting. You can make each piece almost personalized to each user profile, to a point where they feel like you wrote it for them. One company that has been excelling at producing content as part of its distribution strategy is DigitalOcean. With more than 7,000 tutorials, it even created learning paths to help readers better decide what to read and which tutorials to follow. Its content works so well that most search engine top results for relevant questions point to a DigitalOcean tutorial. Each article is well written and also has an appealing aesthetic, making reading it a pleasurable experience.

The way you write and present your documentation has to be aligned with your target audience. Your target audience won't feel engaged if it doesn't find your documentation attractive. Knowing the taste of your potential customers is key to understanding how to craft your documentation so that it resonates with them. One factor in this process is identifying the technical expertise of your readers. Readers who have a beginner level of expertise will find articles with a high technical difficulty hard to understand. On the other hand, advanced readers might find beginner-level articles too simple and boring. The second aspect has to do with the language you use. Independently of the level of expertise of your audience, keep your language simple and easy to understand. In particular, try to avoid any jargon and complex explanations that might only add confusion to the content. The third point has to do with the flow of information. Write your articles in a way that's easy to read and digest. Use bullet points and follow a logical structure to keep readers engaged while going through the information. Finally, try testing different approaches with real users. You can do that manually by interacting with customers and asking for feedback or by performing A/B tests and determining which version works better. A/B tests let you measure the success of two versions of the same content to identify which one works better. Popular tools such as Google Analytics let you run A/B tests and also identify the steps users take from their first interaction to signing up for your API product.

Knowing what happens between the first time potential consumers read your documentation and when they sign up is fundamental to improving your distribution strategy. This type of analysis is called *conversion path tracking* and can produce information that lets you understand the percentage of total readers that end up converting into API consumers by signing up. It also shows you the steps through the conversion path and the percentage of readers that give up on each step. You can see a conversion path as a funnel that gets thinner as potential users go through it. At the top of the funnel is one of the articles you wrote explaining how to solve a specific use case. At the very bottom of the funnel is your successful signed-up user. The more steps you have in between, the higher the chances of abandonment are. Your task is to put yourself in the shoes of potential users and create as frictionless a path as possible.

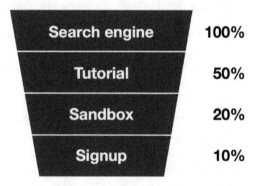

Figure 15.2 – An example of a conversion path with the percentage of users at each step

One way of removing friction from the signup process is by offering an API sandbox. A sandbox lets potential users see how the API works before signing up. It increases the chances of a successful signup because it makes users feel more confident in the API product after seeing it work. A sandbox is even more powerful if combined with a specific example that solves the use case that the user is trying to solve. You can, for instance, have a tutorial link to a sandbox pre-filled with code that uses the API to solve the mentioned use case. Before a user signs up, the sandbox can make requests to a mock API server. The goal is for the user to see how the API works, not to get real results. After signup, users can use the sandbox against the real API and obtain real results. The goal is to get users to learn how your API can solve their problems, see the code to make a request to your API, see how the request to the API would work, and finally sign up to be able to use the real API.

Summary

Right now, you know the role that distribution channels play in attracting customers to your API product. You know what a GTM strategy is and how you can use customer personas to identify your target market. You also know how to craft a pricing strategy by calculating the CAC and making sure that it's always lower than the CLTV. You understand how to use an API portal as a way to make your API easy to discover by consumers. You also know how to engage existing users in a community as a way to share knowledge and build a distribution channel. You understand the power of API marketplaces and how you can use them to your advantage. Finally, you know how you can use documentation as a way to disseminate information about your API and open a door to new customers. More importantly, you know the relationship between these different tools as a way to effectively increase your user base.

The chapter started by showing you what API distribution is and how it's related to a GTM strategy. You learned that the best way to attract users is to be available where they are. You also learned how identifying your target market is key to deploying an effective distribution strategy. You also learned that using customer personas is a good way to identify your potential customers. You learned that there's a cost for every newly acquired user. This is called the CAC. You then learned how you can combine different monetization models to create a pricing strategy that adapts to the expectations of your customers. After that, you learned about API portals, communities, marketplaces, and documentation distribution channels. You learned how to build your API portal in a way that attracts – and retains – users. You also learned that you should focus on use cases and tutorials and not so much on API reference documents. You then learned how to encourage your existing users to collaborate in a community and share knowledge. You learned that a community can be a powerful tool to attract new customers. After that, you learned how API marketplaces work and how you can get your API listed on one. You then learned how documentation, if done properly, can be an effective way to spread information about your API product. You learned what a conversion path is and how to make it as frictionless as possible. Finally, you learned that a sandbox can increase the confidence users have in your API before they even sign up.

These are some of the topics you learned in this chapter:

- Distribution is the process of making your API available to potential customers

- An effective distribution strategy makes the API available on the tools consumers already use

- One critical aspect of distribution is a GTM strategy

- The goal of GTM is to generate revenue by attracting customers

- The size of your target market can dictate the success of your GTM strategy

- By using customer personas, you can understand your target market

- The CAC measures the cost of acquiring each new customer

- You can combine different monetization models into a pricing strategy that aligns with customers' expectations

- The API portal works as the storefront of your API product and has the responsibility of attracting users
- You should focus on writing documentation that shows users how they can solve their use cases
- Having a vibrant community is a way to share knowledge and attract more customers
- Marketplaces attract API consumers for the convenience they offer during the process of discovering an API
- The power of API marketplaces lies in their ability to offer a full end-to-end experience to consumers
- The way you write and present your documentation has to be aligned with your target audience
- One way of removing friction from the signup process is by offering an API sandbox

At this point, you know how to use distribution channels to promote your API product and attract new customers. You understand the power of understanding customer personas and adapting your approach to match the needs of your target market. Getting new customers is crucial for the commercial success of your API product. However, keeping your existing customers satisfied is also mandatory; otherwise, they will quit. In the next chapter, you'll learn how to support existing users to get the most out of your API.

Part 5:
Maintaining an API Product

This part of the book focuses on user-centric aspects of API management, starting with strategies to ensure user success, encompassing support channels, forums, and prioritization of user feedback. It then delves into the challenges of managing multiple API versions, addressing breaking changes and adopting machine-readable methods for effective communication with users. The part concludes by addressing the sensitive topic of API retirement, covering its definition, considerations, user communication, and the importance of conducting retrospectives to document valuable insights gained from the retirement process.

In this part, you'll find the following chapters:

- *Chapter 16, User Support*
- *Chapter 17, API Versioning*
- *Chapter 18, Planning API Retirement*

16
User Support

Even the users of the best API products in the world will, at some point, need support to get their jobs done. Interacting with user support doesn't mean the API product is not well built or the quality is not the best. Instead, it's often a sign that the API needs to evolve beyond how it was designed in the first place.

You'll begin this chapter by learning how you can help users get their jobs done. You'll learn that user support is a set of resources that you can use to help users whenever they have problems. You'll then learn about the benefits of a good user support program. Not only that, but you'll see that users will feel increased trust in your API product if it has good user support. You'll also learn that having users requesting your assistance means they're engaged. You'll learn that you can use all the feedback you get from users to enhance your API product. You'll then see how to create support channels that empower users to find answers to their concerns. You'll learn about different support channels such as a knowledge base, email, and live chat. You'll also learn that a good support system is a combination of multiple channels. You'll learn how to prioritize the feedback you get into bug fixes and feature requests. You'll learn that ticketing systems let you communicate with your internal product and engineering teams. You'll then learn about prioritization methods such as MoSCoW, Kano, and the Eisenhower Matrix. Finally, you'll learn how to create the right measurement practices to understand how your support program is performing and adapt it if needed.

When you finish reading this chapter, you'll know how fundamental good user support is. You'll know that user support is helpful for both you and the consumers of your API. You'll also know the different available support channels and how they can all work together. You'll be able to convert support requests into bug fixes and feature requests. You'll also know the different methods for prioritizing which bug fixes and feature requests you'll consider and work on. Finally, you'll know how to measure your support program to learn what works and what you can improve upon or adapt.

In this chapter, you'll learn about these topics:

- Helping users get their jobs done
- Support channels
- Prioritizing bugs and feature requests

Helping users get their jobs done

In the world of APIs, user support is a set of services, resources, and assistance that you provide to consumers to help them get their jobs done. API support typically involves offering guidance, troubleshooting instructions, and responsive measures to guarantee that users can have the best possible experience. Every time a user feels friction, there's a chance for support to step in and help. The moments when API users feel the need for support the most are during onboarding, any time you introduce a change, and when something stops working as expected. You can offer support by having a dedicated team or by engaging one or more product managers directly with users. In the end, the goal of API support is to address any issues users might have, clarify their doubts, and ensure the best interaction throughout the API life cycle.

From a user perspective, good API support offers the guarantee that, whenever something doesn't go as expected, there's always someone available to interact with and answer questions. API support benefits start right at the onboarding stage. By addressing any difficulties that users might have while signing up, you increase the adoption rate of your API and decrease the time it takes a user to make a first request. By engaging with users when they're sharing their difficulties, you'll be able to understand their specific use case and help them build an integration with your API that has a high chance of being successful. You'll also gain credibility and build trust because users feel they can count on you to help them get their jobs done. Another way to look at user support is to think of it as a source of information for improving your API.

Firstly, whenever users request support, it means they care about your API, and they want to help you succeed in providing them with the best experience. If they didn't care, they'd simply ignore you and move to one of your competitors. What really keeps users engaged and sharing their thoughts with you is their loyalty to your API product. The more customers complain, the more they're interested in seeing your API product succeed. Remember: they're spending their time giving you ideas on how your API can better align with their expectations—for free.

Secondly, you should consider all the information users give you as fundamental to the growth of your API product. Whether it's a report about something not working correctly or a request for a feature that you don't yet offer, every piece of information you get can guide you on how to improve your API. The best way to obtain value from feedback is to ask users to openly share the problems they're having while trying to use your API. Instead of asking for information about specific features or bugs, let users express freely, in their own words, the difficulties they're having. Your job is then to manage the channels that let users communicate their problems and categorize their feedback into actionable information.

Support channels

The best way to provide support is by giving users a way to find the answers to the issues they're having. This type of support channel is called a knowledge base, and its goal is to become the **source of truth** (**SOT**) to all questions that users might have. However, there's a fine balance between what you provide on a knowledge base and how much you invest in interactive support—where real people are involved. Lean too much on a knowledge base, and you lose a sense of what challenges users are having and oftentimes make them feel frustrated when they can't find an answer tailored to their problems. Spend too little on a knowledge base, and you end up repeating the same answers to users who have similar issues.

The ideal scenario is a combination of a knowledge base with other support channels. Let's first see how you can create and grow a productive knowledge base that helps users and also your own support team. The most basic knowledge base you can put together is a web page with common questions and answers. This is often called an FAQ, which stands for "frequently asked questions." You start it with a list of questions that you anticipate users will have, and then you grow it by adding new questions as you get more feedback. Over time, this list will grow and will eventually become hard to navigate. That's when you need to improve it by making it simple for users to find answers to their questions.

Fortunately, there are tools in the market that let you build knowledge-base solutions. A good example is Zendesk, a support tool that lets you build a knowledge base and connect it to other support channels such as email. Connecting the knowledge base system to the email support channel is important because one feeds from the other. During a support session, you're able to pull information from the knowledge base and share it with users. On the other hand, every new insight you find while providing an answer to a user can be fed into the knowledge base. Every interaction with a user becomes an opportunity to increase the information available in your knowledge base.

Another tool that follows a similar approach is Intercom. However, instead of connecting the knowledge base with email, it joins it with the chat support channel. In the case of Intercom, chat support can become automated by obtaining information from the knowledge base. Whenever a user needs information that's not yet available, a real support agent is called into the chat to provide contextual help. The information provided during the chat session can then be fed into the knowledge base, improving it over time. Let's now look at how to use email and chat as support channels.

In online interactions, email is the oldest form of communication and one that's been used as a support channel for the longest time. The biggest advantage of email is its asynchronous nature. Users can share their experience using your product without having to wait for a response from your side. Meanwhile, you have the freedom to investigate the reported problem without having to provide an answer immediately. Email support can also be easily set up and grow as your user base and support team grows. You can easily start offering email support just by having an email account and being able to receive messages and reply to users. As the complexity of the support structure grows, you can use tools that help you coordinate requests among members of a support team and categorize feedback for easier resolution. If email support looks so good, why are there other support channels?

While the asynchronous nature of email support works to your advantage, it's also what makes it less appealing to most users. According to a study conducted by *Techjury*, in 2023, more than 60% of users prefer other support channels over email. The reason is that most users want their problems solved immediately when they reach support. Their expectation, when they have a problem, is that you'll be able to work with them to solve it. Email creates a time buffer where users know they'll have to wait for their questions to be answered, and that generates frustration. The popular alternative that users prefer is live chat. However, setting up a live chat system is not straightforward. While email access is ubiquitous, live chat needs to be set up in a way that works across different systems and device types.

My recommendation is to offer a live chat channel that works with what your audience uses. For example, if most of your audience uses Google, then it's a good idea to use Google Chat for support. Other options include popular chat systems such as WhatsApp from Meta (Facebook) and iMessage from Apple. All these systems offer business-ready chat tools that let you track conversations and have multiple support agents assigned to the same channel. Using one of these systems has several advantages. The first one is you get the whole system already implemented and ready to use. Secondly, your users don't have to learn new skills to interact with you because they can communicate from their preferred tools. And finally, in the case of Google and Facebook, you can offer chat support directly from search results.

If, on the other hand, your audience is scattered in terms of the tools they use, you're better served by setting up a live chat on your website. Here, you have many options, both open source and commercial. The important thing to keep in mind is you'll have to integrate the chat system into your website, and that requires some implementation effort from your side. Rocket.Chat, for instance, is one of the most popular open source solutions and it also offers a commercial solution. On the fully commercial side, Intercom and Help Scout are among the most popular solutions. Whichever tool you decide to use, pay attention to its ability to grow with your needs and to let you categorize and coordinate support requests. Features such as message labelling and support agent assignation are essential to consider as the volume of support requests increases. When you reach a point where your API product becomes popular and many users are requesting support, you want to pay attention to a more public channel: social media.

It's often during times of crisis that customers needing support use the power of social media to make their voices heard. Social media outlets such as LinkedIn, Facebook, or X are powerful channels where users can publicly share their feelings about your API product. Also, because of the technical nature that an API has, pay close attention to other channels such as Stack Overflow and the YCombinator Hacker News. The feelings users share on any of those channels can be positive, but they can also damage the reputation of your business if left unattended. It's then crucial to have a presence in the social networks where your audience spends time to not only evangelize your product but also react to any opinion that users share publicly.

By now, you'll realize that your focus should be not on one single support channel but on multiple touchpoints with your audience. You will probably start with a simple-to-set-up channel such as email and evolve from there. Interactions between users and different support channels will tend to evolve over time. Also, the interactions between support channels, also known as "handoffs," will increase as the amount of requests grows. A support request that originates on LinkedIn might be transported to a private email channel. From there, information from the knowledge base might be used to help resolve the issue, which will be communicated to the user via email:

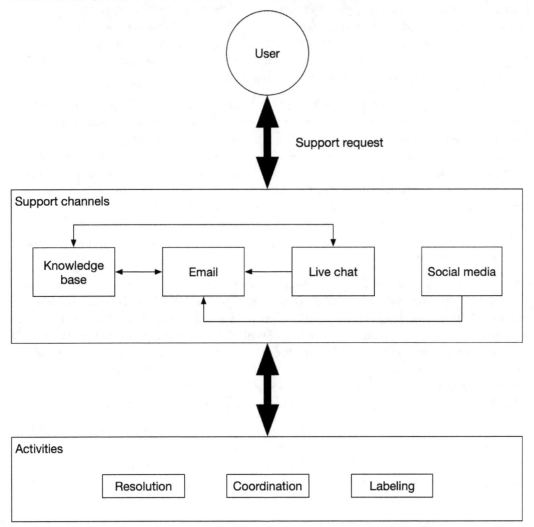

Figure 16.1 – An overview of the interactions between different support channels

All this communication and coordination is very interesting, but unless you have a clear perspective of what is important, you won't be able to solve every user's problems and demands. Let's now see how you can decide what to work on and what to postpone based on how important each request is to your business strategy and objectives.

Prioritizing bugs and feature requests

User support prioritization begins with understanding what type of request a user is making. Every time a user finds a problem and engages with a support agent, the problem can translate into something the user didn't do correctly—which is easy to explain and fixes the problem—or into something that isn't working as expected. Then, the problem can be categorized as a bug or as a feature request. Understanding what the problem is before trying to fix it is crucial. Otherwise, you end up doing everything users are asking and lose track of your own business objectives.

Even though both bugs and non-existent features can be the origin of users' problems, they're quite different. A bug, or defect, is something that happens when a feature of your API behaves in a way that deviates from what was originally designed. By contrast, a feature request is a proposal for a new functionality that isn't currently available on your API. The impact a bug has is negative, and fixing it has the immediate benefit of addressing the problems of all users who are using the affected feature. A feature request, when implemented, results in an impact proportional to the value it creates for the entire user base. However, a feature request doesn't necessarily improve the existing quality of the API product as much as fixing a bug does. So, how do you decide if you should fix the bug immediately or work on a feature request? First, you need to have a ticketing system in place to collect and manage all bug reports and feature requests.

While support channels are the interface between your team and your user base, ticketing systems are what you use to communicate the feedback you gather to your product and engineering teams. And you have to communicate in a way that the internal teams understand. To do that, you first have to try to replicate whatever problems users are complaining about and document the steps you took. Then, you have to categorize the problem as either a bug or a feature request. If it's a bug, you also have to give it a score based on the negative impact it has on the whole API product. If it's a feature request, you give it a score based on your perception of the value it can generate for your user base if it's implemented. Then, you share the ticket you just created with your product team.

A product manager, after looking at an incoming ticket, can quickly assess its importance by understanding if it represents a bug or a feature request and looking at its score. Then, you can use a prioritization framework such as MoSCoW or the Eisenhower Matrix to sort the tickets by order of importance and understand what you need to implement next. Let's see in detail how some popular prioritization methods work:

- **MoSCoW**: The acronym stands for **Must-haves, Should-haves, Could-haves, and Won't-haves**. It categorizes features or tasks into these four priority groups. It helps you distinguish between critical requirements (must-haves) and those that can be deferred (won't-haves).

- **Kano**: Categorizes features into basic needs, performance needs, and "delighters." It assesses customer satisfaction based on how well these different types of needs are met.

- **Eisenhower Matrix** (shown in *Figure 16.2*): Classifies tasks based on their urgency and importance, dividing them into four quadrants: *Urgent and Important, Important but Not Urgent, Urgent but Not Important,* and *Neither Urgent nor Important.*

- **Value versus Complexity matrix**: Assesses features or tasks based on their potential value to the user and the complexity of implementation. It helps prioritize initiatives that offer high value with relatively low complexity, allowing teams to maximize impact while managing resource constraints.

- **RICE scoring**: The acronym stands for **Reach, Impact, Confidence, and Effort**. Each criterion is scored on a scale, and the scores are used to calculate a prioritization score for each feature.

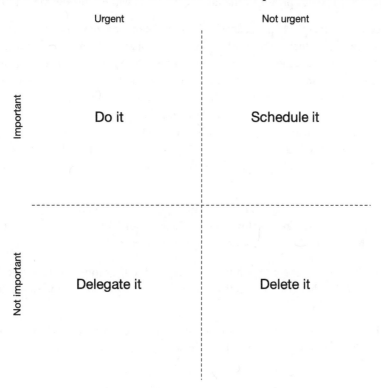

Figure 16.2 – The Eisenhower Matrix gives you a direction of what you
should do with a task based on its importance and urgency

There are other prioritization frameworks, but these should give you a taste of the different approaches you can take. From a user support perspective, you should follow a prioritization methodology that gives importance to customer satisfaction. The Kano model, for instance, focuses on how much each prioritized item contributes to customer satisfaction. You can use other methodologies as long as you use customer satisfaction as a part of your prioritization. For example, with MoSCoW, you can identify must-haves as bug fixes or features that generate a high level of customer satisfaction. With RICE, for instance, you can use the reach and impact criteria to identify items that generate the highest satisfaction for the largest number of users. Whichever approach you take, it's important to understand that priorities are fluid and can change dramatically over time.

The ability to adapt your API development to an ever-changing reality is a must to maintain the quality of your product and keep users satisfied. Doing it is more difficult than it sounds, though. One strategy you can employ is defining features as small as possible and having short development cycles. The key here is being able to add or remove items from your roadmap easily without compromising the whole development effort. Having small features that you can put together like bricks gives you the ability to do a replacement at a low cost from an engineering perspective. By running short development cycles, you can quickly reassess priorities and change direction without having to wait for too long.

As an exercise, consider a fix for a bug that affects a substantial number of users is underway. Suddenly, more users request a feature that makes the bug fix redundant and introduces a better alternative to the previous feature. You can let the development team finish the bug fix and then work on the feature that will make the fix obsolete. Or, you can stop the bug fix and switch the priority to building the new feature. In a situation similar to this one, what's important is how you measure the success of your prioritization. If success for you means that you don't have any bugs, then you should continue with the bug fix. If, on the other hand, success means that your product is evolving to align with users' expectations, then you should prioritize the new feature. In any case, it's easier to switch direction when the development effort is split into small chunks and short cycles. However, the degree to which you're successful with your approach can only be known if you have the right measurements in place.

Let's now see how you can measure the success of your API product support program. To begin with, you should start measuring the **key performance indicators**, or **KPIs**, that matter to the success of your business. Here's a list of the most common support-related KPIs:

- **Response time**: The average time it takes to respond to user inquiries or support tickets. A quick response time indicates a proactive and responsive support team and results in increased customer satisfaction.

- **Resolution time**: The average time it takes to resolve user issues or address support tickets. Efficient issue resolution contributes to user satisfaction and minimizes downtime for API users.

- **User satisfaction**: The overall satisfaction of users based on surveys or feedback collected after support interactions. You can use the **customer satisfaction** (**CSAT**) score to measure the satisfaction of your API consumers.

- **Knowledge-base usage**: Measures how frequently users access and use the knowledge base. A high usage rate indicates that users are empowered to find answers independently.

Altogether, the feedback loop that you create with user support is vital to the continued effort of maintaining and evolving the quality of your API product. Without proper user support, you wouldn't be able to adapt and would risk seeing your customers switch to a better competitor.

Summary

At this point, you know that offering support is a sure way to help your users get their jobs done. You also know that support not only helps customers feel more satisfied, but it also gives you information that helps you enhance your API product. You now know how to set up different support channels such as a knowledge base, email, and live chat. You know how to go from a simple FAQ to a fully blown knowledge base. You also know that the ideal support scenario is a combination of several channels. You know how to translate the information users share into actionable tickets. You also know how to categorize user feedback into either bug reports or feature requests. You know how you can choose which bugs to fix and which features to implement by following a prioritization method. You know about prioritization methods such as MoSCoW, Kano, and RICE. You also know how you can adapt your API development to incoming requests by keeping the workload as small as possible. Finally, you know what the metrics are to put in place to understand how your support system is behaving and improve it, if needed.

You started the chapter with a demonstration of how you can help your API consumers get their jobs done. At the beginning of the chapter, you learned that user support is something you can use to increase the satisfaction of your customers. You learned that the ultimate goal of support is to fix problems users have during their interactions with your API. You then learned that the feedback you receive from support can also help you improve your API. You learned that the best way to get feedback is to let users express themselves freely. You then learned about the knowledge base, email, and live chat support channels. You learned that a knowledge base system can feed other channels by providing answers to recurring questions. You learned that the best support channels are the ones that work with the tools your audience already uses. You then learned about interactions between the different support channels and which tools you can use for each one of them. You also learned that feedback from users can lead to bugs you need to fix or new features you might choose to implement. You learned you can prioritize what to pick based on decision frameworks such as MoSCoW and RICE. Finally, near the end of the chapter, you learned how you can measure the effectiveness of your support system. You learned about metrics to consider, such as the resolution time and CSAT score, which measures the overall satisfaction of users.

Here are a few of the things you learned throughout this chapter:

- User support is a combination of guidance, troubleshooting instructions, and responsive measures
- Users feel the most need for support when they're onboarding, whenever you introduce a change, and when something breaks
- The more users share their feedback, the more they're interested in the success of your API product
- The information users share is fundamental to the growth of your API
- The most popular support channels are the knowledge base, email, and live chat
- You can start a knowledge base by compiling an FAQ
- A knowledge base can be a tool to help you find relevant information when interacting with users

- The advantage of an email support channel is that it's asynchronous and gives you time to provide a response

- However, most users prefer other support channels over email

- Live chat is the most preferred support channel because of its immediacy

- Your focus should be on providing a combination of support channels to cater to the different needs of your audience

- You can prioritize requests coming from users by following decision frameworks such as MoSCoW

- One way to adapt to sudden changes coming from support requests is by working in short development cycles, with chunks of work as small as possible

- You can measure the success of your support system by calculating, among other things, user satisfaction and the time it takes you to resolve incoming requests

Right now, you know how important user support is to the perception users have about your API and also to its growth. It's through user support that you learn directly from users what you must improve and how. Users feel most of the need for user support when things don't work as expected, especially at times when you introduce changes to existing functionality. The next chapter focuses precisely on what to do when you need to introduce changes. Keep reading to learn all about communicating changes to your audience with the help of API versioning.

17
API Versioning

User support feels the most pressure whenever you introduce changes to your API. Whenever users have trouble consuming your API, they'll reach out to support for help. API versioning is a way to anticipate the needs of consumers and inform them that changes are coming. By following a robust versioning and communication strategy, you can make your API easier to use and understand. Continue reading to learn how to apply the best strategy for your API.

This chapter begins by introducing you to the world of multiple API versions. You'll revisit the different versioning strategies that you first learned about in *Chapter 13*. Then, you'll continue learning how to have a default latest version and, at the same time, provide access to earlier revisions. You'll see an example of how GitHub follows a similar strategy. You'll then learn what a deprecation policy is and how to apply one to your API. You'll see how to communicate the deprecation of a version to consumers by crafting and following a sunset policy. You'll also learn how integration tests can help identify incompatibilities between versions. This leads to being able to identify breaking changes, which is something you'll learn next. You'll see how capturing the differences between different versions is crucial to understanding the impact of changes. You'll also learn what you can consider a breaking change and what you can see as innocuous. After that, you'll learn about the different possibilities for communicating changes to your user base. You'll learn what an API changelog is and how you can build one from the list of captured differences. You'll also learn that you can create a machine-readable changelog that can help both you and your consumers in different situations. Finally, you'll see that you can effectively communicate changes using channels such as a community, social media, a blog, and a newsletter.

After reading this chapter, you'll understand that API versioning is much more than having a version number. You'll know that with versioning, you can communicate changes to users, helping them adapt before things stop working. You'll have learned about the types of breaking changes and how they can negatively impact API consumers. You'll have also learned how to create a changelog as a way to record and disseminate information about the changes you introduce to your API. Finally, you'll know how you can use different channels such as your blog or a newsletter to update users every time you modify your changelog.

This chapter covers the following topics:

- Managing multiple API versions
- Breaking changes
- Communicating changes

Technical requirements

In this chapter, you'll explore a command-line tool and a machine-readable document type that uses the JSON format. Expert knowledge in these areas is not required. However, it's helpful to understand the minimum to get the most out of this chapter.

Managing multiple API versions

In *Chapter 13*, you learned about the most common API versioning strategies. Having a versioning strategy means you care about your consumers' expectations. Whenever you change your API version, you're signaling to your users that you've introduced changes. Let's start by reviewing some versioning strategies you can use:

- URL path
- HTTP query parameter
- HTTP version header
- HTTP Accept header
- Message topic or channel

All API versioning strategies have the common goal of helping consumers identify the version they want to interact with. However, they also help you define how you want your API consumers to behave. It's important to identify – and communicate – what you want consumers to do when you announce a new API version. Do you want all your consumers to use the new version immediately? How will they move to the new version depends on how you identify what the "latest version" is on each API request they make.

One strategy that promotes a seamless transition into the latest available API version is offering it by default. Whenever consumers don't indicate which version they want to use, you serve them your latest version. Consumers who want to stick to a particular version can do so by indicating it using the API versioning strategy that you support. As an example, GitHub lets users identify the API version they want to consume using the X-GitHub-Api-Version header. If consumers don't care about which version they're using, they can simply make requests without specifying it. In that case, they'll interact with the latest available version that GitHub has to offer.

However, you still need to have a version deprecation policy. If you're doing API versioning, you can't simply kill all versions except the latest one. Otherwise, users who were consuming a previous version will stop having an API to interact with. Not only should you decide how many versions you'll support, but you also need to determine the longevity of each version. Consumers need time to migrate from a deprecated version to a newer one. It's your job to identify how long that can take and make sure you maintain versions until all existing customers migrate successfully. Also, pay attention that any migration consumers need to do creates friction on their side, and also an opportunity to move to a competitor.

All this being said, you should communicate clear expectations on how long you'll maintain each version and what will happen to consumers of versions that you no longer support. Depending on the nature of your API and existing integrations, consumers might need months to migrate out of deprecated versions. To mitigate this situation and give consumers the information they need, you can define and document what is called an API sunset policy. It's a document that identifies what your company will do whenever an API version enters a state of deprecation. Here are some of the most important elements that you can find on an API version sunset policy:

- **Timeline**: An outline of the important moments to consider during the API sunset. These include the date of deprecation and a transition period where both the deprecated version and a newer one will coexist.

- **Communication plan**: Details on how your team will communicate the sunset to affected users. The plan should include the notification methods you'll use when you send the notifications, and how many times you'll alert users before the sunset happens.

- **Deprecation process**: All the steps that you and your team need to go through to deprecate the API, including sending warning notifications, offering alternative options, and providing technical support during the transition period.

- **End-of-life procedures**: The specific actions you need to take once the API reaches its end of life. These include the termination of any related services, and actions regarding data retention, if applicable.

- **Migration assistance**: The resources and guidance documents to assist affected users in migrating to a newer version of the API or alternative solutions.

- **Support and maintenance**: The level and duration of support and maintenance you'll provide after the sunset happens.

- **Legal information**: Any legal aspects, such as changes to terms of service, liability, and data ownership after the API sunset.

- **Documentation availability**: Information on how affected users can still access the documentation of sunset API. Accessing information is important for comparison with a newer version and guidance during a transition.

- **Compliance**: Details on how the API sunset might affect or interfere with any existing industry standards and regulations.

Even with a sunset policy in place, there will always be users who won't be able to migrate to a newer version in time. Should you make the requests from those affected consumers stop working? Or should you move them automatically to a newer version after the sunset happens? To answer these questions, you need to understand the compatibility between the deprecated version and the new one. One way to find out how compatible different versions are is to have a set of integration tests that you can run against both versions and analyze the results. If all the integration tests pass on both the sunset and the new version, it means there's a high chance both versions are compatible. If, on the other hand, some tests fail, you know what features of the new version aren't compatible with the sunset one.

Knowing what features create incompatibility issues is crucial to providing proper guidance and support to consumers migrating away from the sunset version. You can even create, if possible, a conversion mechanism that makes requests work on both versions during a transition period. The topic of understanding and managing incompatibilities is so important that there's a name for it in the API space: breaking changes. Keep reading to understand what breaking changes are and how you can proactively identify them even before you run any integration tests.

Breaking changes

You can identify changes between different API versions reactively by running integration tests and seeing which features fall. However, that's not the best way to identify what changed – or what will change – between two API versions. There's a proactive approach you can follow even before you write any code. After you re-design your API and update your machine-readable definition document, you can compare it to the one from a previous version. Because the document is machine-readable, it's easy to feed it to a tool that calculates the differences between both versions. This type of tool generates what we call an API **diff**.

API diffs are usually lists of differences between two JSON documents. Here's an example output from `oasdiff`, a popular open source OpenAPI diff tool:

```
1 changes: 1 error, 0 warning, 0 info
error[request-property-became-required] at payments.json
in API POST /payments
the request property 'cardCvv' became required
```

In this case, the tool detected that I made a previously optional property required on the POST / payments operation. This is one type of change that introduces an incompatibility with a previous version. If consumers don't change their requests to add the new required property, they'll get an error as the response. Because of that, you should consider this a breaking change.

Breaking changes are so important that some large API providers explain what they are in their documentation. That's the case of GitHub, which offers extensive information on what you can identify as a breaking change coming from their API. You can view GitHub's list as a good starting point for understanding what kinds of breaking changes there are. oasdiff, the API diff tool you saw in action previously, can detect more than 100 kinds of breaking changes. Let's look at some of these changes in detail:

- **Removing an operation**: If you remove an entire operation, you're creating an incompatibility for consumers who have been consuming it. An operation is a combination of an HTTP verb with a path – for example, POST /payments.

- **Removing or renaming a parameter or a payload property**: By affecting any parameter – or property, if the operation expects a JSON payload – you're making the operation incompatible with how it was before.

- **Removing or renaming a response field**: Similarly, if you change the shape of an operation's response, consumers will have to adapt their code.

- **Adding a new required parameter or payload property**: If you add new required parameters it's obvious that previously working requests will start to fail because they don't have that parameter.

- **Making a previously optional parameter or payload property required**: Similarly, if you make a parameter required, you will affect all consumers who haven't been sending the parameter in their requests.

- **Changing the type of a parameter or response field**: While changing the type might not seem like a big deal, it can introduce crucial incompatibilities on both requests and responses. On a request, parameters are validated by the API server, and if their values aren't aligned with their types, an error is raised. On a response, the types of response fields or properties are used by the consumer code to validate and process the received information. If a type changes, unexpected things can happen on the client side.

- **Removing enum values**: An enum, or enumeration, is a list of constant values. With OpenAPI, you can use enums to specify different values that a request parameter or response property can hold. An example is the definition of a sort query parameter that can hold asc for performing a sorting in ascending order and desc for sorting data in descending order. By removing one of the possible values of a parameter, you're creating potential problems with consumers who have been using it.

- **Adding a new validation rule to an existing parameter**: Whenever you increase the level of validation for any parameter or payload property, you're making it harder for consumers to send their data through. While a change of this nature might not affect most consumers, some of them will see that their requests stop working.

- **Changing authentication or authorization requirements**: These are the most important elements of your API. By making any change to the way you authenticate and authorize consumers, you're risking making all their requests fail. Without a working authentication, consumers won't be able to access your API operations. And, without a working authorization, they won't be able to perform the individual actions they should have access to.

As you can see, there are many ways in which you can break the promise that your API will always be working. While some of them seem innocuous, others are more obvious to spot. Using an API diff tool is mandatory and having integration tests helps you understand what changes you're introducing. However, consumers also need to be aware of any possible incompatibilities. That's why communicating changes effectively is crucial in the process of API versioning.

Communicating changes

Sharing information about API changes with consumers doesn't have to feel like you're sending bad news. Many changes are expected by consumers. Some of these changes are even the result of feature requests that came from consumers. The best way to create the habit of communicating changes whenever you deploy a new API version is by maintaining and updating a changelog. API changelogs are an effective way to share with readers whatever changes you introduced since the previous version.

An API changelog is a document that's usually a part of your documentation. It consists of a list of changes grouped by API version in reverse chronological order. Every time you create a new API version, you can automatically generate a diff that will be included in the changelog. You can generate the diff using the `oasdiff` tool you saw in action previously, or you can use any available commercial service. What's important is that your changelog provides detailed data about each change, along with more contextual information about the state of your API at the moment the change occurred.

For each version you release, there will be a group of changes. At the top of the list of changes, make sure to include detailed release notes explaining why you released a new version and, in your own words, what consumers can expect from it. If you put yourself in the shoes of consumers while writing the release notes, you'll better understand the impact the new version will have on their workflows.

Great release notes start with the key highlights of the new version, summarizing the significant changes and explaining how existing consumers can benefit from the new version. Then, they describe a list of all the new features that were introduced, along with any bug fixes or other enhancements. Great release notes also have a dedicated security section sharing any relevant information. Finally, they include any incompatibility issues that you've identified and explain what users should do if they're affected.

You follow the release notes with a list of the changes, starting with the most important ones: the ones that break existing functionality. You should identify any breaking changes and explain their impact to consumers who have a running integration with your API. Alongside each breaking change, give users enough information for them to mitigate the issue on their side or migrate to the latest version of your API. This is where you can also share any deprecation-related information, including a timeline of the things users should expect to happen.

One way to make each item of an API changelog easy to understand and potentially to be interpreted by software is to adapt the Activity Streams W3C recommendation. Activity Streams is a protocol that lets people describe any activity using machine-readable JSON documents. In the context of the protocol, an "activity" is a way to semantically represent an action that occurred. The Activity Streams model defines a way to describe actors, verbs, objects, targets, and any other element included in an action. As an example, let's translate the breaking change we detected previously with `oasdiff` into a textual Activity Streams entry. It would look something like "Version 1.2.0 updated POST / payments (the request `cardCvv` property became required)." Here, the activity actor is "Version 1.2.0," the verb is "updated," the object is "POST /payments," and there's an explanation of what was updated. This is what the entry could look like in a machine-readable JSON format:

```
{
  "type": "Update",
  "published": "2023-12-09T16:05:14+00:00",
  "summary": "Version 1.2.0 updated POST /payments (the request
property cardCvv became required)."
  "actor": {
    "type": "API",
    "name": "Version 1.2.0"
  },
  "object": {
    "type": "Operation",
    "name": "POST /payments"
  }
}
```

You can use this machine-readable version of the changelog entry for several purposes. The first, most obvious thing you can do with it is produce an HTML output. This would let you fully automate the process of generating a visual list of API changes. In addition, if you provide enough details on the JSON document, you can generate integration tests from the list of changes. By knowing what changes happened and how they affected each of your API's operations, you can make integration tests that verify only what's relevant at the time of the change. Something else you can do is give your consumers access to the machine-readable list of changes and let them use it for whatever they see as interesting. They might even come up with new and interesting ideas.

Having a changelog is crucial to maintaining transparent communication with your API consumers. Additionally, there are other things you can do to increase the level of participation from your audience. As you saw in *Chapter 15*, letting your user community participate can lead to more engagement and the increased success of your API product. Fostering the participation of your API community to discuss each version announcement and suggest improvements increases the transparency and trust consumers have in you. You don't have to build a full community from scratch yourself. The easiest way to get started is to publish your most recent changes where you believe your audience is regularly. Reddit, for instance, is a service that hosts communities of all types. You can identify a few relevant Reddit communities and create a changelog post there. With time and as your level of participation increases, you'll be able to understand what the most interesting communities are.

Similarly, using any type of social media service is also beneficial. The more you spread the information about your new API version, the more chances you'll have of reaching someone who will find value in the information you're sharing. In the case of social media, you can first write a post on your blog that contains all the details about the new API version and a link to the corresponding changelog entry. Then, you can publish a link to the blog post on several social media services. API consumers will have a way of consuming the information inside a space that you control – your blog – where they'll have immediate access to other parts of your website, including documentation, the full API changelog, and technical support.

You should also include the blog post in the next newsletter you send to your whole user base. While sharing a blog post on social media is more immediate than using a newsletter, not all users are constantly on social media. You can think of the newsletter as a more traditional fail-proof way of sharing information with your user base, even if it takes longer for you to disseminate the information. In the end, your goal is to be able to share the changes as quickly and easily as possible with all your API consumers.

Summary

At this point, you know how you can use API versioning to release changes efficiently. You know what the different API versioning strategies are and how consumers can use them to request the version they want to use. You also know that the easiest way to promote an automatic migration to your latest version is to make it the default one. Whenever consumers don't specify which version they want to request, you serve them the latest available one. You know what deprecating an API version means and how you can control its impact by defining a sunset policy. You also know that the highest impact comes from changes that create incompatibilities with the previous version-breaking changes. You know how to identify breaking changes by generating a diff between two API versions. Additionally, you know what types of differences can be considered breaking changes. Finally, you know the options you have for communicating changes to your whole user base. You know that having an up-to-date API changelog is the best way to keep your users updated about all your version releases.

You started by reviewing the API versioning strategies you learned about in *Chapter 13*. You then learned that the strategy that promotes frictionless migration to the latest version is the default one. After, you learned what an API version sunset policy is. You learned about some of its most important elements, such as a timeline, a communication plan, a deprecation process, and migration assistance. You continued by learning about the impact of incompatibility between different API versions. You learned what a breaking change is and how you can detect it by calculating the differences between versions. In practice, you saw an example of using the `oasdiff` open source tool to detect differences and breaking changes. You also learned what the most common breaking changes are and how they affect consumers. Then, you packed the information about versioning and breaking changes to learn how to offer efficient communication with consumers. You learned what an API changelog is and what elements you should put into it. Finally, you reflected on using several channels, such as your blog, a newsletter, and social media, to reach out to your audience and update them every time you release a new API version.

These are some of the topics you learned about in this chapter:

- Users can specify the API version they're interested in using the URL path, an HTTP query parameter, a header, or a message topic for asynchronous APIs

- Making your last API version the default one promotes a seamless transition whenever you introduce changes

- However, you can't simply kill all versions except the latest one

- Migrating from a deprecated version to a newer one takes time, sometimes even months

- You should have an API version sunset policy where you include things such as a timeline, a deprecation process, migration assistance, and documentation availability

- API diffs are a way to detect changes between two versions

- A breaking change is any difference that introduces an incompatibility with the previous API version

- Examples of breaking changes include removing an operation, adding a new required parameter, and changing authentication and authorization requirements

- An API changelog is a list of changes grouped by version in reverse chronological order

- Each entry on an API changelog should begin with release notes for the version being announced

- API changelog entries can be machine-readable documents, making them easy to use with software

- You can announce new API versions using communication channels such as your blog, newsletter, social media, and technical communities

At this point, it's clear that API versioning is fundamental to having a healthy relationship with users. Without versioning, you wouldn't be able to introduce any changes to your API. One part of API versioning has to do with deprecating a single version. But what happens when you need to deprecate the whole API? We'll cover this in the next chapter, so keep reading to learn more.

18

Planning for API Retirement

API retirement refers to the intentional and planned discontinuation or deprecation of a whole API, not just a single version. It involves removing an existing API from service, indicating that it is no longer supported and that existing users should transition to an alternative solution. The suggested alternative solution can be an API from another company or a newer one that's been built to replace the one that's being retired. Whatever the case, you should create a grace period to allow users to successfully migrate out of the retired API. Keep reading to learn how to manage the retirement of your API effectively.

You'll start this chapter by understanding when is the right moment to retire an API. You'll see the key events that can trigger the decision to put an API to rest. You'll see that the decision to start a retirement process comes from factoring in the costs of keeping your API running against the benefits that it generates. You'll learn that retirement is not an end by itself. Instead, it's a process that needs to be managed because it affects the different API stakeholders. As you'll see, one big part of the retirement process is how you communicate it. You'll also revisit the different communication strategies you learned about in *Chapter 17*. In doing so, you'll learn that consumers will need time to adjust to the fact that the API will cease to exist. To mitigate that, you'll see how you can retire your API in phases instead of simply turning it off at once. You'll learn about the different elements of a good retirement communication plan. You'll also learn how to update any existing documentation, listings in API marketplaces, and even ongoing integrations with third-party services. You'll then see what an API product retrospective is and how it can help you avoid making the same mistakes in future APIs. You'll learn that celebrating success is a good way to start the retrospective in a positive mood and avoid blaming. You'll then learn about the people you should invite to participate in the retrospective. You'll see what each type of participant will bring to the table. You'll also learn how to gather quantitative and qualitative data to inform your decisions. Finally, you'll learn how the retrospective ends by defining a list of action items from the learnings.

After reading this chapter, you'll know that API retirement is a process that can take as long as stakeholders need to adapt. You'll understand that the main reason for retiring an API is related to the cost of maintaining it. You'll also know how to replace the API you retire with a new one you build or with an existing competing alternative. You'll also know that communicating the retirement is crucial so that consumers know what to expect. You'll understand that you should use all available channels to share the news in a concerted fashion. Finally, you'll know how to conduct an API product retrospective, invite the right participants, and gather the right information to generate learnings you can use in the future.

In this chapter, you'll learn about the following topics:

- When should you retire an API?
- Communicating API retirement
- API product retrospective

When should you retire an API?

As a general rule, you should put an API out of service whenever the costs of maintaining it are too high compared to what you can gain by keeping it running. There are, however, a few events that can trigger the decision to retire an API. The first one is related to the technology that powers your API. Technology evolves and, sometimes, the one you used to build your API is no longer maintained. When that happens, keeping your API running is risky. If you have an issue, there's no way to obtain support from software that has no maintainers. Additionally, the technology maintainer is no longer available to provide security updates. Keeping the obsolete technology and using it to power your API can become a huge cost for you the longer you postpone getting rid of it. So, it's only natural that you decide to retire that API and build one that replaces it using new technology.

Another reason that makes you ponder if you should retire your API is when all or some of its dependencies no longer exist. In this case, the problem is not with the API's technology, but with the one powering its dependencies. If you continue maintaining your API in this situation, you could get into a situation where you won't be able to update your API because its dependencies are obsolete. Replacing dependencies with alternatives can be an option. However, if the effort of doing it is too high and its outcome too precarious, you're better off building a new API from scratch using the latest dependencies.

Once you've gone a while without updating your API, its quality will certainly suffer. You won't be able to keep up with all the latest security and industry updates, and your consumers will notice. Moreover, they will keep making feature requests and reporting bugs that you can't commit to work on. Your SLA will be repeatedly violated, and you'll feel that you're losing customers to competitors. If you have seen yourself in this situation because you can't evolve your existing API, then it's time to replace it with a new one. Yes, loss of quality is another reason to retire your API. It's worse to offer a bad API than not offering one at all.

One aspect of quality has to do with industry standards. They evolve to adjust to the latest technology and market trends. You should as well. Otherwise, your API will feel outdated, and you won't be able to serve your consumers as you once were able to. A good example is the introduction of architectural styles such as GraphQL and gRPC. If you were using SOAP at that time, you would have probably seen most of your user base move to a competitor – unless you evolved from SOAP into one of its latest replacements. Notwithstanding, adopting industry standards comes with a cost. Again, if you can't support the cost of change, then it's time to retire your API and build a new one from scratch.

If industry standards scare you, compliance with laws and regulations should keep you awake at night. Not complying with a regulation such as HIPAA or GDPR can lead to paying expensive fines or even putting your company out of business. However, compliance can be costly and take time. If you find yourself in a situation where the technology you're using is outdated, updating your API code is complicated, and it feels almost impossible to comply with regulations, then it's time to think about retiring your API.

At this point, you should be seeing a trend in the arguments I'm presenting. In the end, it all comes down to your ability to maintain your API technology so that it can keep up with customer demand, industry standards, and regulations. The cost of ongoing maintenance is what you need to monitor so that you know when is the right time to retire the API. This situation won't happen overnight, which is a good thing, because it will give you time to plan how you'll manage the retirement.

Once you decide that retiring your API is the path you want to take, it's time to decide what to do with the hole that you're creating. One possibility is building a new API to replace it, improving all the metrics that matter to your business while adapting to the latest customer feedback and industry standards and regulations. Another option is to simply stop offering the API product if that's what makes the most sense from a business perspective. Whatever the option you end up taking, you need to communicate your decision to existing API consumers. Read on to learn how to make the retirement announcement in the best possible way.

Communicating API retirement

Sharing that your API is being retired can be done in a way similar to what you learned in *Chapter 17* when you read about API versioning. However, in this case, you're not focused on changes and incompatibilities. Instead, you're focused on conveying to stakeholders that the API will cease to exist. It's important to use a positive tone, communicate ahead of time, and clearly explain the reasons behind the retirement.

You should trigger the communication process as soon as you know you're going to retire the API. This is to give existing consumers and stakeholders enough time before the actual retirement happens. As with incompatible API changes, consumers need time to adapt. However, in the case of a whole API ceasing to exist, customers might need months to adjust. That's why it's critical to start communicating right from the moment you identify there will be a retirement. It will give you a head start so that you won't have to wait months until you have all your technical work done.

One possibility is to perform a "graceful degradation" of your API, right after doing a formal announcement, and while working on the technical tasks leading to the full retirement. A graceful degradation means your API won't stop working immediately. Instead, it will gradually lose quality over time until it stops working completely. Quality loss can be related to performance, or the ability to maintain data consistency. What's essential is that your consumers will have the real perception that the API is losing its ability to fulfill their needs gradually over time. That situation alone will be enough for most consumers to decide to adapt to the new set of circumstances. Something else that helps consumers migrate out of the API that's being retired is understanding the reasons behind the retirement.

Sharing why you're retiring your API is helpful because it makes your consumers understand your position. Customers will most likely empathize with you and even support your decision if the reasons behind it are aligned with their values. Begin by expressing the decision clearly and transparently. Clearly state that the API is being retired and provide a brief overview of the reasons behind this decision. If the reasons are related to external factors, explain how you tried to mitigate the situation but couldn't cope with the costs of adjusting. If, on the other hand, the reasons are related to internal factors, explain why not retiring the API would be a bad decision for your business.

Clearly communicate how the retirement will impact the API consumers. Explain any potential disruptions to service, changes in API behavior, or features that will no longer be available. Provide detailed instructions and guidance on how customers can migrate from the retired API to the recommended replacement if there is one. Include documentation, code samples, and examples to facilitate a smooth transition. At this moment, everything that is mentioning or pointing to the API you're retiring should be updated. That includes any inbound links from other websites.

Among the many places that have connections to your API portal or documentation, marketplaces, and third-party platforms are the ones you can control the most. In the case of marketplaces, you were the one signing up to be listed. So, now, it's your turn to switch off your presence. Removing your API from a marketplace might be as simple as signing in and doing it yourself. However, completely removing your API might not be the best approach. Think of all the people who will still be searching for your API and won't be able to find any information about its retirement. Instead of doing a removal, you can replace your API entry with information about its retirement. Whoever finds your API will land on a marketplace listing with an explanation about the retirement of the API and instructions on how to find a replacement.

In the case of third parties that are integrating with your API, you'll have to get in touch with someone from their side to either turn off the integration or migrate to an alternative. It's best to list all the third-party integrators currently working with your API first, starting with the ones with the highest number of users. Then, go through the list, reach out to the company behind the integration, and communicate the retirement of the API. Offer support to either migrate to a newer version of your API or to an alternative that can provide similar functionality to existing users. At the end of this process, it's good to look back to see what worked, what didn't, and what you can learn. Keep reading to see how you can drive an API product retrospective.

API product retrospective

An API product retrospective is a systematic review and analysis of the development, deployment, and performance of an API product after a specific period or milestone. This retrospective provides an opportunity for the development team, stakeholders, and other relevant parties to reflect on the API's journey, identify successes and challenges, and discuss opportunities for improvement. As with any retrospective, people tend to focus on what went wrong. Instead, I invite you to start by recalling the success of the API.

Look back to when you first introduced the API you're now retiring. Remember why you built the API and what your business objectives were at that time. Did the API help your organization achieve those proposed objectives? To what extent did the API contribute to the success of your company? Answering these questions helps you and the retrospective participants engage in a positive way of thinking. It's natural that the API retirement invokes negative feelings and makes retrospective participants think of why things have gone wrong. That's why starting with a positive mood is so important to make the best of the retrospective.

Evaluating success is one of the purposes of the API product retrospective. The other two areas where participants should focus on are identifying improvements and planning a replacement API. The goal of identifying improvements is to document what the team believes could be changed in the future to achieve a better outcome. Here, your goal is to identify situations and activities that led to the decision to do the API retirement. You don't have to come up with any solutions at this point. Instead, just document what you think could be done in a better way. Finally, evaluate how planning for a replacement was managed and what its outcome was. Evaluate criteria such as how long the planning took, how the plan was communicated to consumers, and how many people were involved in the planning activities.

Speaking of people, you should define a list of all the participants in the retrospective. The list of key stakeholders can change depending on the situation, but these are the groups of people most common to find in a product retrospective:

- **Engineers**: One or more people representing the engineers who helped build the API product.

- **Product managers**: The ones responsible for defining the API features and overall strategy. Product managers have a broad view of the API because they're the ones aligning business objectives with engineering decisions.

- **QA**: One or more people involved in the quality assurance of the API product.

- **Customer support**: Someone knowledgeable of the interactions between existing API consumers and the product team.

- **Business leaders**: The ones that have participated in decision-making processes leading to the creation, evolution, and later retirement of the API.

- **Customers**: A group of people representing different customer cohorts. This should be a small group of three to five people who can be representative enough to speak on behalf of all the customers.

With a list of participants, you can prepare an invitation. However, before that, you should define the time frame of the retrospective. All participants need to understand the context of the retrospective. One thing is to share thoughts about the whole lifetime of the API; another thing is to focus on just the last month. Be clear about the time frame to set the right expectations. You can also define multiple periods that you identify as important to debate. Whatever the time frame you come up with, the group should be able to use it to obtain the information you're seeking.

To help the team come to conclusions, you must have all the relevant data and metrics available. You should obtain and organize summarized usage analytics from the time frame you're inspecting. By knowing how consumers interacted with the API over time, you'll be able to understand how well their expectations were being met. If there were a lot of errors, for instance, you would know the API wasn't addressing the needs of consumers correctly. The number of errors, along with their type and how often they occur, is the best way to understand what was wrong, if anything, with the API product. However, even if there are no noticeable errors, you can infer a lot by knowing the aggregate response times of the API's most used features. Low response times usually mean that consumers are happy. On the other hand, high response times, often leading to unavailability, mean the API requires improvements. Go back to *Chapter 14* and grab any metrics you feel are relevant to the quantitative part of the analysis.

Another area you should pay attention to is the qualitative understanding of how consumers feel about your API. User feedback is something you can use during the retrospective. However, you can't use all the consumer feedback you have received over time. The best approach is the one you explored in *Chapter 16*, where you categorize user feedback to turn the raw information into actionable metrics. Here are some feedback categories for you to explore during the retrospective:

- **Functionality**: Comments on the features and capabilities of the API. Feature requests and bug reports.

- **API user experience**: Feedback on how easy it is to use the API and suggestions on how to improve the experience.

- **Performance**: Opinions on how fast the API responds to requests and how reliable the responses are over time.

- **Integration**: Feedback on how easy it is for integrators to make the API work with third-party tools and systems.

- **Security**: Concerns related to the security features of the API.

- **Customer support**: Feedback on how fast and effective support was to fix issues.

- **Value**: Perspective from customers on the pricing of the API product and what its perceived value is.

- **Overall satisfaction**: General sentiment from consumers about their satisfaction with the API product, taking into account all the previous factors.

Now, combine the quantitative data you obtained from metrics with the qualitative information you gathered from categorized user feedback. You'll surely find patterns and a connection between both types of information. Start by looking at positive signs and then move to more challenging and problematic areas. Address one key aspect at a time, take note of each of the challenges you faced, and identify lessons to learn from each situation. Finally, discuss possible ways of preventing the same issues from happening in the future by bringing different perspectives from all the participants.

The API product retrospective ends with a list of action items and an intention to follow the recommendations to improve the success of the business. If you're building a new API to replace the retired one, following the retrospective's suggestions is a good idea. If not, you can extrapolate the action items to other APIs you'll build and even to other areas of your company with similar challenges. Make sure you share all the documentation that was created during the retrospective so that all the stakeholders can learn from it and consult it in the future.

Summary

At this point, you know what an API retirement process looks like. You know that retiring an API isn't just turning it off. Instead, it might take as long as existing consumers need to adjust. You know why APIs are retired and how you can measure the factors that can influence your decision. You also know that understanding the cost behind maintaining your API is crucial to knowing when the right time for a retirement is. And knowing the right time is the first thing you need to communicate the API retirement. You know that you should begin by sharing why you're retiring your API and giving plenty of time for all stakeholders to digest the information. You also know how to update existing information about your API in marketplaces to give new and existing visitors the news about its retirement. Additionally, you know how to drive an API product retrospective to learn what led to the retirement and improve your future actions. You know who to invite to the retrospective and how to obtain relevant metrics to inform your decisions. Overall, you know that retirement is just one of the stages of an API that can lead to designing a new API and repeating the life cycle.

This chapter began by giving you the information you need to decide when to retire your API. You learned about the factors that trigger an API retirement and how long the process can take. You also learned that existing consumers need time to adapt to the reality of not having your API available any longer. You then learned how to communicate the retirement to all the API stakeholders, including integrators and marketplaces. You learned that offering support is crucial to offering a smooth API retirement transition. You also learned that you can offer a replacement by building a new API yourself or by suggesting an existing alternative. After that, you learned how to look back at what led to the retirement by doing an API product retrospective. You learned who you should invite to the retrospective and how to gather the necessary information to obtain meaningful learnings. Finally, you learned that retirement is one stage of your API life cycle and that you shouldn't see it as the end.

Instead, you should treat it as an opportunity to design a new and better API and re-engage with existing consumers.

Here are a few of the things you learned in this chapter:

- API retirement is the deprecation of a whole API

- Retiring an API is a process, not an end by itself

- You should retire an API when the costs of maintaining it outweigh its benefits

- Among other factors, retiring an API comes from the difficulty in updating the technology that powers it

- You can replace a retired API with a new one you build or with an existing alternative from the market

- Existing consumers might need months to adjust to the retirement of your API

- You can perform a graceful degradation of your API as a way to retire it gradually

- Communicating the API retirement starts with explaining the reasons behind it to existing consumers

- You should update all the places where your API is listed or integrated with information about its retirement

- Doing an API product retrospective offers a chance to reflect on the journey leading to retirement

- Evaluating success is one of the goals of an API product retrospective

- Engineers, product managers, business leaders, and customers should participate in an API product retrospective

- You can inform the API product retrospective using both quantitative and qualitative metrics

- The outcome of the API product retrospective is a list of lessons and action items

Right now, you have all the knowledge you need to build an API product and even retire it if you have to. However, you shouldn't see API retirement as the end. Instead, it's an opportunity to take what you've learned and go back to designing a better API. Revisit *Chapter 4* to refresh your memory on what the API life cycle is, and go back to *Chapter 5* to start a new API product from scratch. Even though this book ends here, you can always use it as a reference during the design, implementation, release, and maintenance phases of any API.

Index

`Packtpub.com`

Subscribe to our online digital library for full access to over 7,000 books and videos, as well as industry leading tools to help you plan your personal development and advance your career. For more information, please visit our website.

Why subscribe?

- Spend less time learning and more time coding with practical eBooks and Videos from over 4,000 industry professionals

- Improve your learning with Skill Plans built especially for you

- Get a free eBook or video every month

- Fully searchable for easy access to vital information

- Copy and paste, print, and bookmark content

Did you know that Packt offers eBook versions of every book published, with PDF and ePub files available? You can upgrade to the eBook version at `packtpub.com` and as a print book customer, you are entitled to a discount on the eBook copy. Get in touch with us at `customercare@packtpub.com` for more details.

At `www.packtpub.com`, you can also read a collection of free technical articles, sign up for a range of free newsletters, and receive exclusive discounts and offers on Packt books and eBooks.

Other Books You May Enjoy

If you enjoyed this book, you may be interested in these other books by Packt:

API Testing and Development with Postman

Dave Westerveld

ISBN: 978-1-80056-920-1

- Find out what is involved in effective API testing
- Use data-driven testing in Postman to create scalable API tests
- Understand what a well-designed API looks like
- Become well-versed with API terminology, including the different types of APIs
- Get to grips with performing functional and non-functional testing of an API
- Discover how to use industry standards such as OpenAPI and mocking in Postman

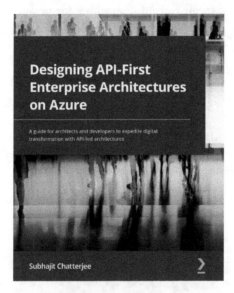

Designing API-First Enterprise Architectures on Azure

Subhajit Chatterjee

ISBN: 978-1-80181-391-4

- Explore the benefits of API-led architecture in an enterprise
- Build highly reliable and resilient, cloud-based, API-centric solutions
- Plan technical initiatives based on Well-Architected Framework principles
- Get to grips with the productization and management of your API assets for value creation
- Design high-scale enterprise integration platforms on the Azure cloud
- Study the important principles and practices that apply to cloud-based API architectures

Packt is searching for authors like you

If you're interested in becoming an author for Packt, please visit `authors.packtpub.com` and apply today. We have worked with thousands of developers and tech professionals, just like you, to help them share their insight with the global tech community. You can make a general application, apply for a specific hot topic that we are recruiting an author for, or submit your own idea.

Share Your Thoughts

Now you've finished *Building an API Product*, we'd love to hear your thoughts! Scan the QR code below to go straight to the Amazon review page for this book and share your feedback or leave a review on the site that you purchased it from.

https://packt.link/r/1837630445

Your review is important to us and the tech community and will help us make sure we're delivering excellent quality content.

Download a free PDF copy of this book

Thanks for purchasing this book!

Do you like to read on the go but are unable to carry your print books everywhere?

Is your eBook purchase not compatible with the device of your choice?

Don't worry, now with every Packt book you get a DRM-free PDF version of that book at no cost.

Read anywhere, any place, on any device. Search, copy, and paste code from your favorite technical books directly into your application.

The perks don't stop there, you can get exclusive access to discounts, newsletters, and great free content in your inbox daily

Follow these simple steps to get the benefits:

1. Scan the QR code or visit the link below

https://packt.link/free-ebook/9781837630448

2. Submit your proof of purchase
3. That's it! We'll send your free PDF and other benefits to your email directly